BIRDS

OF HAWAII, NEW ZEALAND, AND THE CENTRAL AND WEST PACIFIC

Written and illustrated by
BER VAN PERLO

Princeton University Press
Princeton and Oxford

PRINCETON ✔ ILLUSTRATED CHECKLISTS

Published in the United States, Canada, and the Philippine Islands by
Princeton University Press, 41 William Street, Princeton, NJ 08540

nathist.princeton.edu

Originally published in English by HarperCollins Publishers Ltd. under the title:
Birds of New Zealand, Hawaii and the Central and West Pacific

First published in 2011

Text copyright © Ber van Perlo 2011
Illustrations copyright © Ber van Perlo 2011

Library of Congress Control Number 2010942940
ISBN 978-0-691-15188-5

Edited and designed by D & N Publishing, Baydon, Wiltshire
Colour reproduction by Dot Gradations
Printed and bound in Hong Kong by Printing Express

10 9 8 7 6 5 4 3 2 1

CONTENTS

PREFACE

This book should be regarded and treated as a field guide in which the necessary information, needed to identify a bird at the moment you observe it, is given in a condensed form. The low weight and small size thus achieved make it easy to carry the book around and consult in the field. Support of the identification of difficult species can be obtained in more detailed, regional bird books, which could be consulted at home, in vehicles, at your hotel, etc. Sound recordings from commercial CDs and DVDs and transferred to your iPod or other MP3 player can be an important supplementary aid in the field.

It is said that the painting in my books is 'a bit sketchy, somewhat fast and loose, not finely finished'. However, if you see a bird in the field you see its uniform-coloured plumage parts as a single surface; painting each individual feather will give too much information unless the feathers form a pattern. I also find it difficult to draw straight lines, for example, when depicting the parallel primaries in a folded wing, or perfect circles when forming an eye, but this does not prevent a species from being quickly recognised. To me, it is essential that one can see that my work is hand-painted; I love the magic of small spots and streaks, applied with skill and luck that create a shape of something that looks quite different in reality when seen from nearby.

In the introduction I have paid attention to landscape and habitats, specific to the area, that determine which bird species can be seen; that is also why information is given about plate tectonics, as one of the main landscape-forming factors.

An effort was made to update this book to 2009, but a recent record could not be inserted, namely that of American Avocet *Recurvirostris avocetta* on Maui, Hawaii. Very similar to 39.9 but with black mantle and white scapulars.

In June 2010, just before finishing writing this book, the fourth edition of the *Checklist of the Birds of New Zealand* was published by the Checklist Committee Ornithological Society of New Zealand. Only the common names from this publication could be incorporated in this book as explained in the first paragraph of 'Systematics and Names'.

ACKNOWLEDGEMENTS

I owe many thanks to Bas van Balen, who very kindly reviewed the first draft and made many valuable comments. I thank also Valerie Notenboom, who read parts of the text and helped with suggestions.

Many people in the area welcomed me into their homes or were my guides and gave me all sorts of assistance and support. I am very grateful to them; I remember the people in Kaikoura, Kapiti Island and Tiritiri Matangi Island who made it possible for me to see albatrosses and many endemic species; my hosts Ron and Ruth Wilkinson and Lee Marie and Isao; my guide Mick Peryer who showed me around in the Waikanae Estuary and my guide Vido Senivalati who made it possible to see and hear the birds in the Colo-I-Suva Forest. I thank Kiniviliami Ravinaloa for accepting me in his community. I especially want to thank Phylis Gandi, my bird pal on Viti Levu, who made my stay there one of the best experiences in my life.

The skins of rare birds in the British Museum of Natural History were a fine source of information and I very much appreciate the support and assistance of the staff, especially of Mark Adams, Hein van Grouw, Robert Prys-Jones and Katrina Cook.

I am also grateful to Myles Archibald, Associate Publisher at HarperCollins, who made the production of this work possible and to Julia Koppitz, Senior Editor for editing and help provided.

This work was done single-handedly, but I owe everything to the artists and writers who are my predecessors in creating field guides. They are the giants on whose shoulders I stand, or better, in whose shadow I work.

I could not produce a work such as this without so many bird photographers uploading a rich flow of photos to the Internet; it is unbelievable what they produce with their cameras, while I also admire their skills using the Internet.

I enjoy the warm interest of my brothers and sisters, my children Maarten, Annemarieke, Joris and Susan, my grandchildren Charlie, Jake, Mila, Jayden and Terrence and of course my extended family, Thijs, Kick and Dewi, Bart, Jasha and Otis.

But most of all I thank Riet for her patience and support.

Ber
1 June 2010

SYMBOLS, ABBREVIATIONS AND GLOSSARY

Symbols

✳   Habitat: a set of environmental factors that is preferred by a bird
♪   Vocalisation
⊙   Notes on range

Abbreviations

1st W	The plumage worn by a bird after moulting from juvenile plumage
2nd W	The plumage worn in the 2nd winter of a bird's life
♂	Male
♂ ♂	Males
♀	Female
♀ ♀	Females
Ad.	Adult
Ads	Adults
Br	Breeding
N-br	Non-Breeding
C	Central
Cf.	Compare to
E	East(ern)
E.	Endemic
Esp.	Especially
Excl.	Excluding
Extr.	Extreme
I	Introduced
I.	Island
Irr	Irregular visitor
Is	Islands
Imm.	Immature
Imms	Immatures
Incl.	Including
Juv.	Juvenile
Juvs	Juveniles
L	Length in cm
N	North(ern), etc.; also in combination with E and W
NI	New Zealand's North Island
Nom.	Nominate; the subspecies of a species that was the first described
Pl	Plate
R	Rare
S	South(ern)
Sec	Second or seconds
SI	New Zealand's South Island
Ssp	Subspecies
Ssps	> 1 subspecies
Thr.	Throughout
V	Vagrant
W	Wingspan in cm; can also mean West(ern): depends on context

Countries

A.Sa	American Samoa
Co	Cook Islands
Fi	Fiji
FrPo	French Polynesia
Gu	Guam
Ha	Hawaii
Ki	Kiribati
Ma	Marshall Islands
Mi	Micronesia
Na	Nauru
Ni	Niue
NMa	Northern Marianas
NZ	New Zealand
Pa	Palau
Pi	Pitcairn Islands
Sa	Samoa
Tok	Tokelau
Ton	Tonga
Tu	Tuvalu
WaF	Wallis and Futuna

Glossary

Endemic A species that occurs only in an area with well-defined boundaries such as a continent, a country, an island or a habitat.

Forest A tall, multi-layered habitat in which high trees dominate the canopy, which is continuous and closed.

Gallery forest A riverine belt that is dominated by trees.

Jizz Typical silhouette and stance of a bird species.

Riverine belt Any growth along a river or stream that is higher and greener than the more distant surroundings.

Rufescent Tinged with red-rufous colour.

Savanna In this book: any (large) area with a continuous cover of (high) grasses, interrupted by shrub and (5–30%) tree canopy.

Second growth New natural forest developing in places where the original forest has disappeared.

Speculum Patch of colour on the wing contrasting with that of the rest of the wing.

Transient plumage Stage between Br and N-br plumages with traces of Br plumage still visable.

Woodland A habitat in which trees dominate, but the canopy is not closed.

PARTS OF A BIRD

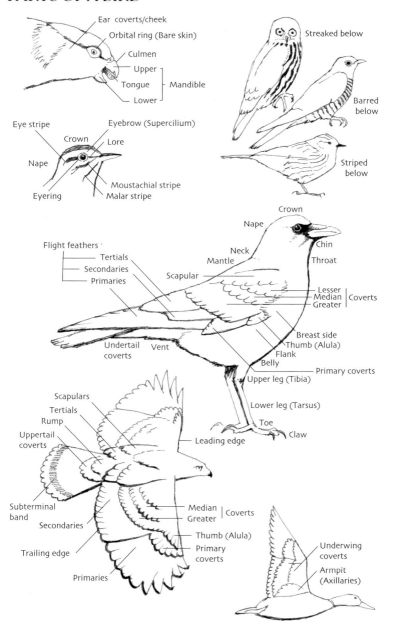

Ear coverts/cheek
Orbital ring (Bare skin)
Culmen
Upper
Tongue
Lower
Mandible

Streaked below

Barred below

Striped below

Eye stripe
Eyebrow (Supercilium)
Crown
Lore
Nape
Eyering
Moustachial stripe
Malar stripe

Crown
Nape
Neck
Mantle
Scapular
Chin
Throat

Flight feathers
Tertials
Secondaries
Primaries

Lesser
Median
Greater
Coverts

Undertail coverts
Vent
Breast side
Thumb (Alula)
Flank
Belly
Primary coverts
Upper leg (Tibia)
Lower leg (Tarsus)
Toe
Claw

Scapulars
Tertials
Rump
Uppertail coverts
Leading edge

Subterminal band
Secondaries
Trailing edge
Primaries
Median
Greater
Coverts
Thumb (Alula)
Primary coverts

Underwing coverts
Armpit (Axillaries)

11

INTRODUCTION

Systematics and Names

In order to apply a uniform systematic approach and set of names that are valid in New Zealand, Hawaii, the USA, Asia, Europe and anywhere else, this book follows James F. Clements *The Clements Checklist of the Birds of the World*, 6th Edition (Helm, 2007) with regard to systematic status (species or subspecies) and names (scientific and English).

However, the best-known or most widely used common names used by New Zealanders and in Hawaii should not be neglected, if only to make it easier to find the birds of this book in regional publications. Therefore, alternative English names for **species**, advised by the Ornithological Society of New Zealand (OSNZ) for New Zealand and by the American Ornithological Union (AOU) for Hawaii, are indicated as insertions in the 'Clements name', preceded by 'NZ' or 'AOU' in superscript. For example:

the species indicated by number 20.9 (Plate 20, number 9) in this book is named the '**INTERMEDIATE** (or [NZ]Plumed) **EGRET**'; the 'Clements name' is **INTERMEDIATE EGRET**, while the 'OSNZ name' is **PLUMED EGRET**;
number 28.1 in this book is named the '([NZ]Australian) **WHITE-EYED DUCK**'; the 'Clements name' is **WHITE-EYED DUCK** and the 'OSNZ name' is **AUSTRALIAN WHITE-EYED DUCK**.

In a few cases, well-known alternative names are inserted in the 'Clements names', such as:

41.5 SNOWY (or Kentish) **PLOVER**

There are also many local names for **subspecies**, mentioned by the OSNZ and in use for those that occur in New Zealand; these names are given in numbered notes following the captions. An example is:

38.5 PURPLE SWAMPHEN[44] *Porphyrio porphyrio*
ssp *melanotus* [NZ]Pukeko;
ssp *pelewensis* [Palau]
ssp *samoensis* [Samoa and Fiji]

In this example, the subspecies Pukeko occurs solely and exclusively in New Zealand (with the extra complication that it is treated by the OSNZ as Nominate of SOUTH-WEST PACIFIC SWAMPHEN [NZ]*Porphyrio melanotus*). In the following example there are four subspecies found in New Zealand:

14.8 LITTLE SHEARWATER[19] *Puffinus assimilis*
Nom. [NZ]Norfolk Island Little Shearwater
ssp *kermadecensis* [NZ]Kermadec Little Shearwater
ssp *haurakensis* [NZ]North Island Little Shearwater
ssp *elegans* [NZ]Subantarctic Little Shearwater

Of these, *elegans* can also be seen outside the area covered by this book, but only in NZ it is known as Subantarctic Little Shearwater. (Note: The OSNZ has recently risen the Clements subspecies *elegans* to independent species [NZ]SUBANTARCTIC LITTLE SHEARWATER *Puffinus elegans*.)

The French names are those of the Commission Internationale des noms français d'oiseaux (CINFO 1993).

The sequence of families in this book is more or less traditional, but strongly adapted in order to include up to a maximum of nine, rarely ten similar-looking species, sometimes from different families, in one plate.

Format of the Species Accounts

Plates

In general, the plumages depicted on the plates are Br. plumages unless otherwise indicated. Normally, the birds on any given plate are painted to the same scale except flight silhouettes, which are normally shown smaller. If ♂ and ♀ have different visual features, such as colouring or dimensions of body parts (e.g. tail length), both are illustrated, unless the differences are only small (e.g. a slightly duller colour of the ♀). Juveniles, immatures and/or first winter plumages are shown when they are often seen in these plumages. N-br plumages are given if the birds visit the area in this plumage; migrants such as waders, which are mainly seen in N-br plumage, are illustrated in this plumage on the plates. If several subspecies of a species occur in the area, and they are distinguishable, these are in many but not all cases illustrated, but not for example, those separable mainly on basis of range.

An effort has been made to show the birds in their typical 'jizz'; what birders call jizz is a difficult to define combination of size, relative proportions and body carriage of a bird. Part of a bird's jizz can be, for example, its stance (the angle of its body axis to the horizontal).

Captions

The information for each species is given in this order:

- *the English name* in bold capitals (with alternative names in standard lower case);
- *the French name* between square brackets;
- *the scientific name* in italics;
- *length* in cm, measured from tip of bill to tip of tail (L) or between the tips of spread wings (W);
- *identification notes* with emphasis on the main features, or those that are not visible in the plates (e.g. the colour pattern that appears when a bird opens its wings) or those that are most important for separation from similar species. Notes on behaviour are often added when important for identification;
- *habitat* (the set of environmental factors, preferred by a bird species) preceded by the symbol ✱; only simple terms are used such as forest, woodland, marsh, plantations, savanna;
- *voice*, preceded by the symbol ♬; where possible a distinction has been made between 'call' and 'song' as being the basic vocalisations of birds, 'call' being any short, probably unrestrainable sound, given by a bird to indicate its presence to himself or other animals, 'song' being the modus in which a ♂ (sometimes also the ♀) advertises its possession of a territory or its mood (anger, nervousness, contentment, togetherness).

In the *voice*, attention is paid to:

- pitch, using a subjective scale 'very low, low, mid-high, high, very high, very/extremely high, extremely high', wherein 'very low' and 'extremely high' indicate vocalisations that are just not quite too high or too low to be audible and 'mid-high' for the normal pitch of an average person's voice, if trying to imitate the vocalisation;
- speed, for which terms are used as 'very slow, slow, calm, rapid, hurried, fast';
- tempo, as defined by the 'length' between notes, indicated by the use or absence of hyphens between notes, so *beep beep beep* is slower than *beep-beep-beep*, while *beepbeepbeep* is the fastest. Also terms such as rattle and trill are used to describe tempo; an apostrophe as in *t'sreee* is used to indicate a short, yet noticeable separation between two consonants;
- loudness, described as soft, weak, loud, ringing, etc.;
- structure, indicated by terms such as accelerated, lowered, gliding, crescendo, staccato, etc.;
- length of song or call given in seconds;
- quality in terms such as harsh, shrieking, mewing, etc.;
- transcriptions – this is the most difficult way to describe 'voice' for several reasons: different

people will transcribe bird vocalisations using different vowels and consonants (for example, compare the way in which in several bird guides the chirping of a house sparrow is transcribed) and the differences that exist between written and spoken text in different languages (French people will transcribe a sound in a different way to a Dutch or English speaker).

To keep transcriptions short, use is made of the punctuation mark '-' to indicate repetition of one or more foregoing notes. In 'preep -' the note(s) are repeated 1–3 times, in 'preep - -' the repetition is more than three times and in 'preep—' the repetition is given very fast.

The use of an acute accent on a vowel, for example in weétjer, means that part is accentuated; the diacritic grave, as in rèh-rèh rèh, is used to indicate that the 'e' sounds as the 'e' in 'red'; similarly, ò sounds as the 'o' in 'pot'; parts of transcriptions written in capitals are uttered louder.

The vocalisations in this book might support an identification or may make it easier to remember a bird sound when it is heard again. However, it should be kept in mind that though the vocalisations of many seabirds, herons and other large waders are described, they vocalise only or mainly in, above or near their breeding colonies.

The description of voice in this book is based on tapes, CDs and DVDs (see Bibliography).

| OCCURRENCE | | | | KEY TO DISTRIBUTION MAPS |
Resident	April–September	October–March	Not season related	
■	■	■		Range
	✳	✳	✳	Frequent visitor
+	+	+	+	Vagrant or rare irregular visitor
◌	◌	◌		Found on or near many, most or all island within enclosure
●				Island population (Indicator ◀)
○				Seabird breeding colony
			⤴	Transient
			?	Hypothetical, possibly extinct or unproven record*
E.NZ				Endemic (New Zealand, etc.)
Int				Introduced Species
			R	Rare in the area, rare in range, irregular visitor or vagrant

There are three levels of maps in this book; a general map of the area (see following page), country maps for each of the 20 political entities (see country maps on pages 24–39) and small distribution maps for each species next to its caption opposite its plate. Abbreviated country names at the end of the captions refer to the country maps, the numbers following the abbreviations refer to the numbered islands on the country maps.
* A '?' in the text, preceding a country abbreviation or an island number, refers to an hypothetical or unproven record or to the possibility that the species is extinct .

Distribution Maps
Information about range, seasonality and occurrence can be an important aid in supporting or weakening an identification. In the distribution maps, attention is paid to these factors, expressed in the key on page 14.

The Area Covered
The area encompasses the following 20 political entities, given in random sequence:

- **Hawaii** (US State)
- **Fiji** (Republic)
- **Tonga** (Kingdom)
- **Nauru** (Republic)
- **Samoa** (Republic)
- **American Samoas** (US unincorporated Territory)
- **Kiribati** (Republic)
- **Marshall Islands** (Republic)
- **Federated States of Micronesia** (free association with the USA)
- **Tuvalu** (British Commonwealth)
- **Tokelau** (New Zealand Territory)

- **Niue** (free association with New Zealand)
- **Cook Islands** (free association with New Zealand)
- **Guam** (US unincorporated Territory)
- **Wallis and Futuna** (French Overseas Collectivity)
- **Pitcairn Islands** (UK Overseas Territory)
- **Northern Marianas** (Commonwealth in union with USA)
- **Palau** (Republic)
- **French Polynesia** (French Overseas Territory)
- **New Zealand** (British Commonwealth)

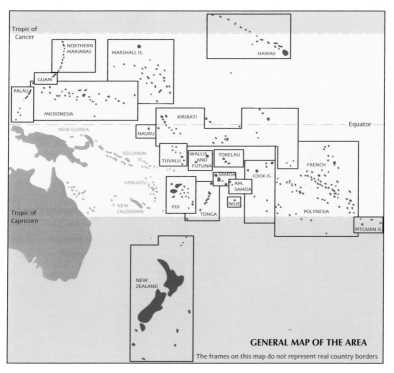

GENERAL MAP OF THE AREA
The frames on this map do not represent real country borders

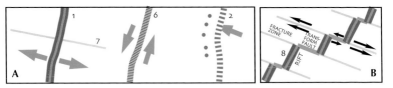

The Environment

Some Basics for Oceania

The following paragraphs give a short introduction to the factors that determine the presence and distribution of bird species in the area. The main factor that dominates all others is that the area is composed of islands, varying between very large and very small, lying either far apart from or close to each other. All these islands have been uplifted from the ocean bottom by tectonic activity; those in the tropics are modified by reef building. Almost all islands, except the most southern ones, were once covered by forest. The final stage in the formation of habitats was the way in which people transformed and used the environment.

TECTONICS The outer mantle of the earth is formed by solid rock (the lithosphere), covered by an accumulation of sediments, volcanic products and changed basic rock (the crust). The lithosphere overlays the asthenosphere, a mantle of plastic flowing rock.

The lithosphere is horizontally subdivided into seven or eight major plates and many minor plates, which ride on the asthenosphere. Some of these plates and parts of them are denser and heavier, lay lower and form the floor of the oceans. The plates move in relation to each other:

- at spreading (divergent) boundaries (A1);
- at collision (convergent) boundaries (A2); and
- at transform boundaries (A6), where two plates move in opposite directions.

The area covered by this book is dominated by a convergent border between the oceanic Pacific Plate and the continental Australian and Filipino Plates (see map 'PLATE TECTONICS').

Tectonic plates separate (or diverge) from each other along a 80,000km long, mainly mid-ocean network (A1) that encompasses the earth. Nearest to the area is a network segment along the west coasts of North and South America. A typical spreading (or divergent) zone (C) can be described as a pair of parallel ridges on both sides of a rift. The rift bottom fills itself with upwelling, red-hot magma, which drives the plates apart and forms new ocean floor.

Collision boundaries are zones of subduction, where heavier oceanic plates dive under lighter continental plates as shown in A2 and D. These zones are marked by a deep trench (D3). When the crust, together with lithospheric material, sinks into the asthenosphere it is heated to such a high temperature that magma chambers (D4) are formed, which float to the surface forming rows of volcanoes (D5) arranged in island arcs (the Kermadecs and Northern Marianas are typical island arcs). These arcs form a sort of perforation, along which the edge of the overlaying plate is often torn off and dragged under itself on the back of the submerging plate.

Other types of conflicting boundaries are also possible (A6), for example, where plates or plate fragments rub along each other under a sharp corner.

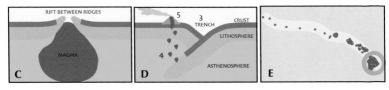

The speed of spreading is unevenly dispersed along the mid-ocean ridge system. Tensions are solved by many fissures (A7) perpendicular to the rift. The rift segments (B8) shift in relation to each other; the parts of fissures between rift segments are called transform faults. The movement at their sides is in opposite directions, which may cause volcanic activity. The outer parts of the fissures are called fracture zones; these separate areas moving in the same direction, which causes no or only low volcanic activity.

Here and there, far from the edges, magma penetrates through the ocean plate. These places are known as hotspots (E); hot magma wells up via these holes giving birth to volcanic islands at the surface. Because the ocean plate moves in a north-westerly direction the hot spots keep drilling holes, forming chains of islands, the youngest being the most eastern one. The Hawaiian islands are a good example.

The map 'PLATE TECTONICS' also shows a transform fault (green line on map), running from the Nazca Plate (near South America) via the Pitcairns to the Line Islands, which could have produced the many islands of the Pitcairn, Tuamotu and Line Islands. However, their origin could also have been a hot spot near Easter Island.

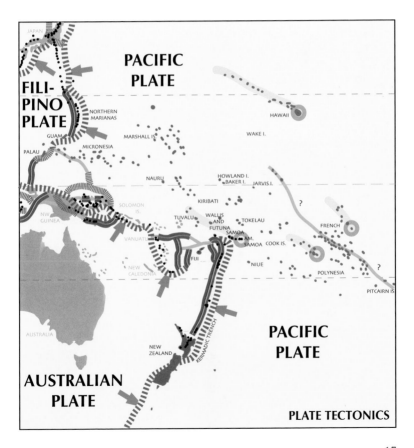

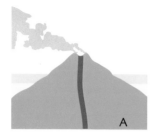

REEF BUILDING There are many species of coral organisms. The group that can build a reef is only found:

- in clear salt waters;
- at depths shallower than 50m (beneath this depths the coral skeletons change to coral limestone, darker yellow-green in figures);
- with an optimum temperature of 26–27°C; and
- strong currents and/or heavy agitation (otherwise food particles are unable to reach the tentacles of the polyps).

Most if not all tropical Pacific islands have a volcanic origin. Reef building starts as soon as a new volcano has emerged (A) and coral larvae have been carried in by ocean currents.

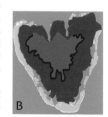

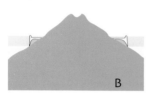

The first stage is a fringing reef (B) at a short distance from land and normally encompassing a shallow lagoon. It takes about 10,000 years from stage A to reach stage B.

In the course of time the volcano erodes or the local ocean bottom subsides. If coral growth can keep up with the speed of this process, a barrier reef (C) is formed on the base of coral limestone; note the wider, locally deeper lagoon.

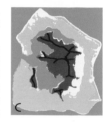

Ultimately, all land will be eroded, and the barrier reef will become an atoll (D) enclosing an open lagoon. The process from A to D can take 30 million years.

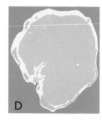

It is possible that an atoll can be uplifted by movements in the earth's crust, by which an uplifted coral island is formed. A limestone rock emerges (E) and becomes encircled by a fringing reef.

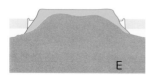

HAWAII

Dry Forest
Mountain Scrub
Alpine Desert

Rain Forest
Agriculture
Lava streams since 1800
Lowland Scrub

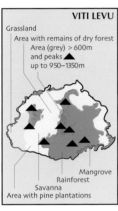

VITI LEVU

Grassland
Area with remains of dry forest
Area (grey) > 600m
and peaks ▲
up to 950–1350m

Mangrove
Rainforest
Savanna
Area with pine plantations

LAND USE AND VEGETATION TYPES Once most islands were mainly covered by forest. The arrival of man brought about many changes in this environment and the following are the main present-day habitats.

Ocean

Open Tropical Ocean: warm water contains less prey (fish, squid, etc.) than cold water, therefore most seabirds in the tropical ocean are more numerous at places where deep cold currents from higher latitudes well up above (under) sea mounts (submerged volcanoes) and at the western edges of the Pacific.

Temperate Ocean: water temperatures between 10 and 18°C, found between the tropics and 48°S and N. Rich in oxygen and nutrients, very rich in fish (less species than in tropical seas, but often in large shoals) and in other life forms.

Coastal habitats

Lagoon: shallow, clear water rich in food for terns, gulls, noddies, tropicbirds and frigatebirds.

Seashore: especially important for migrating shorebirds.

Mangrove: mangrove stands support many bird species and form a habitat where heronries are often found.

Littoral Forest: the forests and thickets bordering the beach.

Lowland forest types

Lowland Dry Forest: found at the dry north-western side of high mountain chains, where the rain, brought in by the eastern trade winds, is released on the eastern slopes. All forms of dry forest are almost completely transformed to agricultural use. In the mixed exotic/native remains a few of the original endemic bird species may be found, plus many alien species.

Lowland Rainforest: as highland rainforest but with a more diverse range of tree species, denser undergrowth and many tree ferns.

Agricultural habitats

Coconut/Breadfruit Forest: found in the coastal areas of many Pacific islands. Mixed with species such as guavas, mango and *Ficus*. This is an ancient man-made habitat.

19

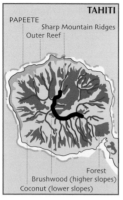

TAHITI
PAPEETE
Sharp Mountain Ridges
Outer Reef

Forest
Brushwood (higher slopes)
Coconut (lower slopes)

Farmland: food crops, fruit orchards, floriculture, vanilla, etc.

Savanna: low production grassland with some tree cover, many breadfruit shrubs and dominated by exotic grasses. Often replaces (dry) forest after repeated burning.

Grassland: areas dominated by grasses with little tree and shrub cover, also replacing former forest. Savanna and grassland in Pacific islands are normally the result of human activity.

Wetlands

Wetlands: rare freshwater habitat in the Pacific; most original wetland is drained and changed to crop- and grassland. Wetland bird species are now dependent on man-made ponds, reservoirs, sewage fields, etc.

Upland forest types

Production Forest: mainly Caribbean Pine or Eucalyptus plantations.

Upland Dry Forest: once covered about one-third of the larger Fijian islands and also was common at the leeside of the Hawaiian islands; now greatly altered to savanna with sparse vegetation.

Montane Rainforest: various forest types united by high humidity and limited temperature variations. Exact timing of dry season varies. Characterised by epiphytes and mosses. This habitat has often disappeared from the smaller islands and the remains on larger islands are threatened.

Cloud Forest: the highest parts of rainforest, which are characterised by a high incidence of fog.

Secondary Forest: new natural forest where the original forest has disappeared. As a habitat it is highly variable, from low woodland to tall forest with more open canopy than virgin forests and lacking old emergent trees.

PALMERSTON ISLAND
(COOK IS)

Bare Beach
Coral Reef, inundated at high tide
Palmforest often with seams of scrub

Other habitat types

Lava Plains and other bare ground at high altitudes: for some bird species this forms an important habitat (Hawaii Goose, Omao, Tahiti Petrel, White-tailed Tropicbird).

GEOLOGY The core of New Zealand was pushed, compressed and folded up against the Australian area some 370 million years ago. About 300 million years later (or 70 million years ago) New Zealand and Australia were separated along a rift that created the Tasman Sea. The rift 'healed' and 25 million years ago the eroded and flattened remains started to be uplifted again.

South Island is dominated by a row of Alps over the full western length. The subduction processes in the trenches north and south of South Island are contrary to each other, pressing the alpine area together. Along the main Alpine Fault both 'Alp-halves' are sliding along each other (the western 'half' moving faster north). Secondary faults are forming the highlands near Kaikoura. This is also the place where the deep Kermadec Trench brings cold, fish-rich water near the coast, attracting a host of seabirds.

North Island is dominated by volcanic activity. The Pacific Plate dives under the Australian Plate that carries the island. Where the plate sinks into the liquid-hot asthenosphere, magma is released and 'floats' to the surface forming an arc of volcanoes in the Taupo Volcanic Zone. In this zone and elsewhere there are several lakes in places where once large to very large magma chambers exploded (each one forming a 'caldera'), after which the remains collapsed and were filled by rain water. It is also the zone where many hot springs are found; the water that surfaces here (often as steam) is heated deeper down in the earth's crust. The volcanoes and calderas outside the Taupo Volcanic zone are mainly remnants of older volcanic arcs.

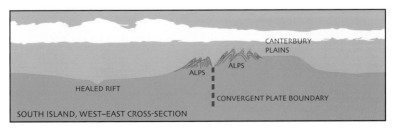

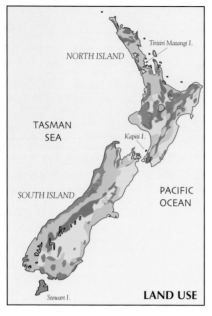

Tiritiri Matangi I.

TASMAN
SEA

Kapiti I.

SOUTH ISLAND

PACIFIC
OCEAN

Stewart I.

LAND USE

LAND USE AND VEGETATION TYPES

Originally the greater part of New Zealand was covered by forest. After the arrival of the Polynesians (AD1000) and later on the settlement of the Europeans (from about AD1840) more than 50% of the forest was cleared, mainly by fire and grazing. Originally there were no mammals in New Zealand except seals and three bat species, but when the first people arrived they introduced unwillingly or with full intent a wide range of animals, which started to compete with or to live on the native flora and fauna, driving many endemics to complete or near-extinction. The most dangerous intruders were black rats, mice, feral cats and possums. The latter not only eat eggs and nestlings of indigenous species but also the saplings and fresh shoots of indigenous trees and plants. The best places to see remaining endemics are a few islands and a single mainland sanctuary that have been made predator-free.

Before the arrival of the Europeans there were very few deciduous tree species, the 13m-high Tree Fuchsia (*Fuchsia excorticata*) among them.

The main habitats include the following:

Farmland, orchards, high producing grassland: together covering about 24% of the total land area.

Extensive grasslands: low vegetation, mainly with exotic and indigenous grasses. Livestock tend to be grazed over large areas. Some extensive grassland may have conservation or recreational uses.

Mosaic of different types of forest and farmland: among these is coastal forest, which is not very tall and is made up of plants that can tolerate salty winds. It generally lacks large conifers and has fewer vines and epiphytes. The canopy is dense and wind-shorn.

Lowland forest: resembles tropical rainforest, but is less rich in species. Found mainly in the northern half of North Island. A typical species is Kauri, a gigantic conifer with small oval leaves. Further south other tall species such as Southern Beech dominate. In most places all or most tall trees have been felled.

Upland forest: dominated by Southern Beech. Undergrowth less dense than in lowland forest. Before the introduction of possum, deer and goat, was rich in berry-producing shrubs.

Planted forest: mainly pine species, but also eucalyptus. Often with rich undergrowth of native plants. Covers about 7% of total land area.

Lakes: the lakes of South Island are drowned glacier valleys, those of North Island are mainly water-filled calderas (collapsed magma chambers).

Note: most natural inland wetlands of New Zealand have been drained for agriculture.

The Birds

Island Avifauna

The area covered by this book is characterised by its immense water surface, its countless small islands and the often enormous distances between them. This makes the avifauna typically island in character.

Of the 125, worldwide recognised true seabirds (albatross and petrels), 87 (or 70%) occur in the area.

As a result of the islands never having been connected to other land masses, the birds that are present have arrived on their own wings or with sea currents, by evolution or by human introduction (Polynesians and Europeans). About 70 species (9%) have been introduced, at least 201 species (25%) are endemic, so about 510 species came unsupported from elsewhere.

Before the arrival of humans there were no mammals in the area, except seals, bats and fruit-bats. There were not many birds of prey either. Many bird species could therefore afford to lose the ability to fly; 12 flightless species (excluding penguins) can still be found, namely five kiwis (all species), three rails (Weka, Takaha and Henderson Island Crake), three ducks (Auckland Duck, Auckland Teal and Campbell Teal) and one parrot (Kakapo).

The arrival of man with his domesticated animals and stowaways, such as black rats and the wilful introduction of predators such as weasels and stoats (as means to control other introduced animals such as rabbits), meant the extinction of many species. Fairly recently it appeared that avian diseases spread by mosquitoes have been responsible for the extinction and near-extinction of several species on the Hawaiian Islands (e.g. the Hawaiian Crow). In the mid 1960s the Brown Tree Snake sneaked into Guam. Without the natural predators of the snake or adequate defence by the autochthonic birds the Guam Flycatcher was extirpated and the Guam Rail could only survive in captivity.

An appendix to this book presents a list of 59 species that have become extinct since 1800. The islands' isolation has also had these effects:

- many species are fearless and docile. Indigenous birds in New Zealand, for instance, are often easily approachable;
- the genus *Acrocephalus* spread over many islands and split up in 12 endemic species; these differ mainly in the degree of leucism. The Tahiti species occurs also as a rare melanistic morph;
- islands tend to produce dwarf (none in the area) or giant species. The giants are extinct (unless one wants to consider Takahe, the world's largest rail and Kakapo, the world's heaviest parrot, as such). Extinct are the ten ostrich-like, forest dwelling Mao species in New Zealand (the largest reached a length of 2.7m). They were preyed upon by the likewise extinct Haast's Eagle, which, with a length of 1.7m, was the largest raptor that lived in historical times. Other giants – likewise extinct – were Moa-nalos (giant ducks from Hawaii), Vitilevu Giant Pigeon and Eyles's Harrier from New Zealand;
- a large amount of endemics. An endemic is a species that occurs only in an area with well-defined boundaries, such as a continent, a country, an island or a habitat. In this book 201 species are mentioned as endemic; only those, occurring with their full life-cycle solely in one of the 20 countries of the area, are treated as endemics, but does not include those restricted to, for example, a group of countries or the whole area. Subspecies are incidentally mentioned as endemic in the captions.

The Endemic Species of the Area

The following pages show maps and the endemics as thumbnails (not depicted to relative scale), arranged per country. The numbers preceding the species refer to the plates and numbers of the species on the plates, while the numbers at the end of the entries refer to the islands on the maps of the countries where the species occur.

Hawaii Endemics

304 species, including the following 33 endemics:

11.6 **Hawaiian Petrel** . Breeds Ha 1,2,4,?5,?7
    Pterodroma sandwichensis
24.6 **Hawaiian Goose** . . . . Ha 1,2,5,7
    Branta sandvicensis
27.2 **Hawaiian Duck** . . . . Ha 1,2,6,7,8
    Anas wyvilliana
27.4 **Laysan Duck** . . . . Ha 13,17
    Anas laysanensis
32.6 **Hawaiian Hawk** . . . . Ha 1
    Buteo solitarius
38.7 **Hawaiian Coot** . . . . Ha 1–7
    Fulica alai
74.1 **Omao** . . . . Ha 1
    Myadestes obscurus
74.2 **Kamao** . . . . ?Ha 7
    Myadestes myadestinus
74.3 **Puaiohi** . . . . Ha 7
    Myadestes palmeri
74.4 **Olomao** . Ha ?5 (no records since 1980)
    Myadestes lanaiensis
75.8 **Millerbird** . . . . Ha 8
    Acrocephalus familiaris
78.8 **Elepaio** . . . . Ha 1,6,7
    Chasiempis sandwichensis
87.1 **Maui Parrotbill** . . . . Ha 2
    Pseudonestor xanthophrys
87.2 **Akikiki** . . . . Ha 7
    Oreomystis bairdi

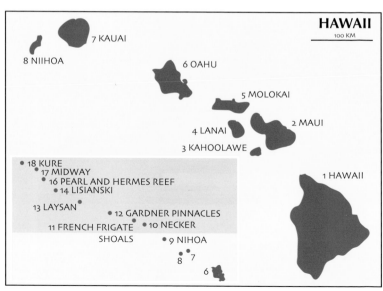

25

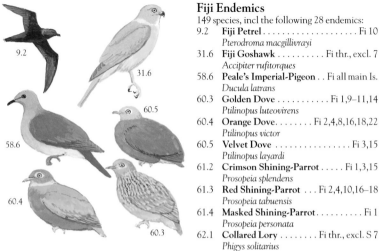

Fiji Endemics

149 species, incl the following 28 endemics:

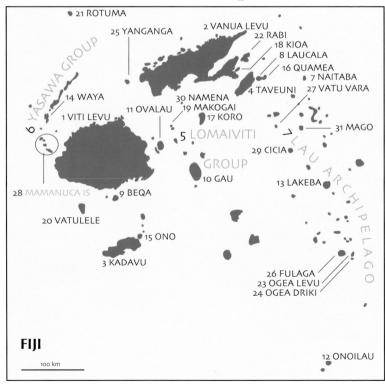

21 ROTUMA

25 YANGANGA

2 VANUA LEVU

22 RABI

18 KIOA

8 LAUCALA

16 QUAMEA

7 NAITABA

27 VATU VARA

YASAWA GROUP

14 WAYA

11 OVALAU

30 NAMENA

19 MAKOGAI

17 KORO

4 TAVEUNI

1 VITI LEVU

5 LOMAIVITI

31 MAGO

LAU ARCHIPELAGO

29 CICIA

GROUP

10 GAU

13 LAKEBA

28 MAMANUCA IS

9 BEQA

20 VATULELE

15 ONO

3 KADAVU

26 FULAGA

23 OGEA LEVU

24 OGEA DRIKI

FIJI

100 km

12 ONOILAU

27

Tonga Endemics

73 species, including the following 2 endemics:

34.3 **Niuafo'ou Scrubfowl**.........Ton 7
　　　Megapodius pritchardii
84.4 **Tongan Whistler**..............Ton
　　　Pachycephala jacquinoti

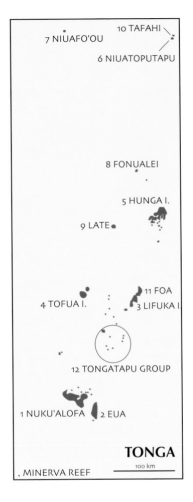

7 NIUAFO'OU

10 TAFAHI

6 NIUATOPUTAPU

8 FONUALEI

5 HUNGA I.

9 LATE

4 TOFUA I.

11 FOA

3 LIFUKA I.

12 TONGATAPU GROUP

1 NUKU'ALOFA　　2 EUA

TONGA

100 km

MINERVA REEF

76.9

Nauru Endemics

27 species, including the following endemic:

76.9 **Nauru Reed Warbler**............Na
　　　Acrocephalus resei

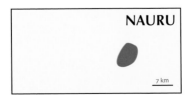

NAURU

7 km

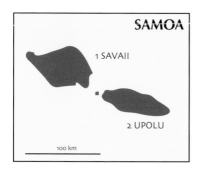

Samoa Endemics

82 species, including the following 8 endemics:

57.2 **Tooth-Billed Pigeon** . . . . Sa
Didunculus strigirostris
67.5 **Flat-Billed Kingfisher** . . . . Sa
Todiramphus recurvirostris
72.3 **Samoan Triller** . . . . Sa
Lalage sharpei
79.6 **Samoan Flycatcher** . . . . Sa
Myiagra albiventris
80.6 **Samoan Fantail** . . . . Sa
Rhipidura nebulosa
83.1 **Samoan White-Eye** . . . . Sa 1
Zosterops samoensis
84.6 **Samoan Whistler** . . . . Sa
Pachycephala flavifrons
86.8 **Mao** . . . . Sa
Gymnomyza samoensis

American Samoa

66 species, no endemics.

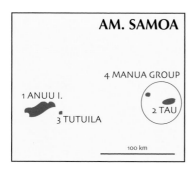

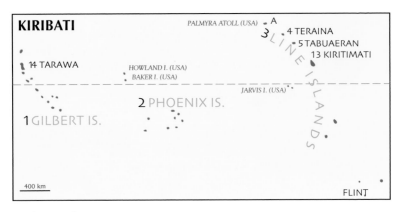

KIRIBATI

PALMYRA ATOLL (USA) A
3 4 TERAINA
5 TABUAERAN
13 KIRITIMATI

14 TARAWA
HOWLAND I. (USA)
BAKER I. (USA)

JARVIS I. (USA)

2 PHOENIX IS.

1 GILBERT IS.

L I N E   I S L A N D S

400 km

FLINT

Kiribati Endemics

74 species, including the following endemic:

76.3 **Christmas Island Warbler** . . . Ki 4,5,13
Acrocephalus aequinoctialis

Marshall Islands

83 species, no endemics.

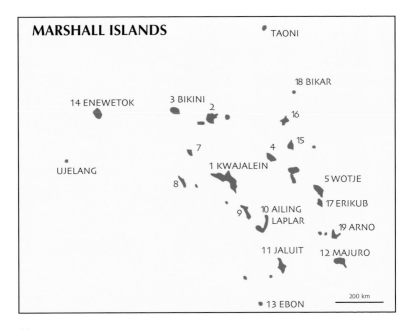

MARSHALL ISLANDS

TAONI

18 BIKAR

14 ENEWETOK   3 BIKINI
2
16

7   4   15

UJELANG   1 KWAJALEIN

8   5 WOTJE

17 ERIKUB

9   10 AILING   19 ARNO
LAPLAR

11 JALUIT   12 MAJURO

200 km

13 EBON

MICRONESIA

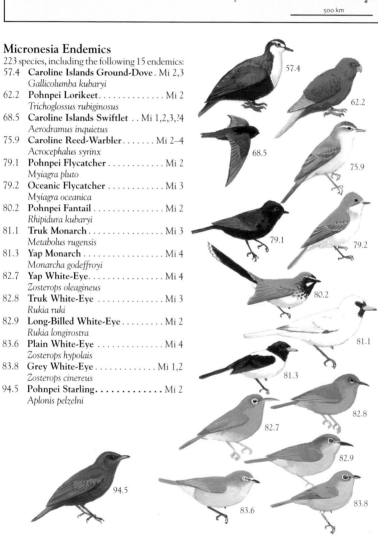

4 YAP

2 POHNPEI

3 CHUUK (or TRUK)

1 KOSRAE

500 km

Micronesia Endemics

223 species, including the following 15 endemics:

57.4 **Caroline Islands Ground-Dove** . Mi 2,3
  Gallicolumba kubaryi
62.2 **Pohnpei Lorikeet** . . . . Mi 2
  Trichoglossus rubiginosus
68.5 **Caroline Islands Swiftlet** . . Mi 1,2,3,?4
  Aerodramus inquietus
75.9 **Caroline Reed-Warbler** . . . . . . Mi 2–4
  Acrocephalus syrinx
79.1 **Pohnpei Flycatcher** . . . . Mi 2
  Myiagra pluto
79.2 **Oceanic Flycatcher** . . . . Mi 3
  Myiagra oceanica
80.2 **Pohnpei Fantail** . . . . Mi 2
  Rhipidura kubaryi
81.1 **Truk Monarch** . . . . Mi 3
  Metabolus rugensis
81.3 **Yap Monarch** . . . . Mi 4
  Monarcha godeffroyi
82.7 **Yap White-Eye** . . . . Mi 4
  Zosterops oleagineus
82.8 **Truk White-Eye** . . . . Mi 3
  Rukia ruki
82.9 **Long-Billed White-Eye** . . . . Mi 2
  Rukia longirostra
83.6 **Plain White-Eye** . . . . Mi 4
  Zosterops hypolais
83.8 **Grey White-Eye** . . . . Mi 1,2
  Zosterops cinereus
94.5 **Pohnpei Starling** . . . . Mi 2
  Aplonis pelzelni

31

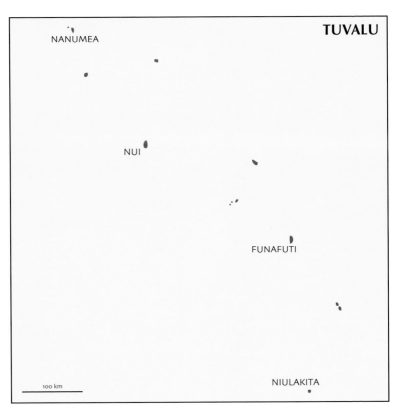

Tuvalu
34 species, no endemics.

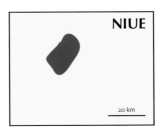

Tokelau
29 species, no endemics.

Niue
31 species, no endemics.

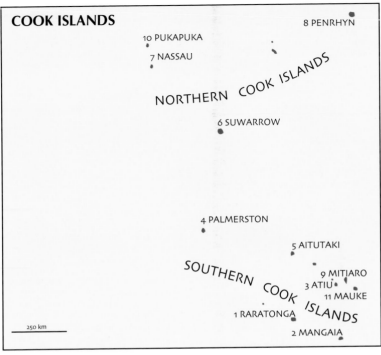

COOK ISLANDS

8 PENRHYN

10 PUKAPUKA

7 NASSAU

NORTHERN COOK ISLANDS

6 SUWARROW

4 PALMERSTON

5 AITUTAKI

SOUTHERN COOK ISLANDS

9 MITIARO
3 ATIU
11 MAUKE

1 RARATONGA

2 MANGAIA

250 km

Cook Islands Endemics

50 species, including the following 6 endemics:

59.2 **Cook Islands Fruit-Dove** . . . . . . Co 1,3
     Ptilinopus rarotongensis
67.2 **Mangaia Kingfisher** . . . . Co 2
     Todiramphus ruficollaris
68.4 **Atiu Swiftlet** . . . . Co 3
     Aerodramus sawtelli
76.5 **Cook Islands Reed-Warbler** . . . Co 2,9
     Acrocephalus kerearako
78.1 **Rarotonga Monarch** . . . . Co 1,3
     Pomarea dimidiata
94.2 **Rarotonga Starling** . . . . Co 1
     Aplonis cinerascens

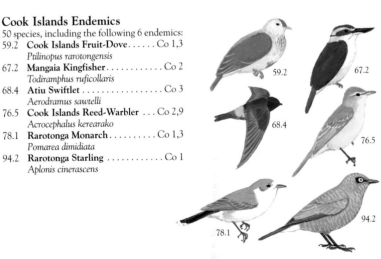

GUAM

50 km

WALLIS AND FUTUNA

1 FUTUNA
2 ALOFI

3 WALLIS

100 km

PITCAIRN ISLANDS

4 OENO
1 HENDERSON I.
2 PITCAIRN I.
3 DUCIE

100 km

36.6

Guam Endemic

93 species, including the following endemic:
36.6 **Guam Rail** . . . . Gu, NMa 4(I)
     Gallirallus owstoni

Wallis and Futuna

39 species, no endemics.

37.4

59.8

62.6

76.6

76.8

Pitcairn Islands Endemics

43 species, including the following 5 endemics:
37.4 **Henderson Island Crake** . . . . Pi 1
     Porzana atra
59.8 **Henderson Island Fruit-Dove** . . . . Pi 1
     Ptilinopus insularis
62.6 **Stephen's Lorikeet** . . . . Pi 1
     Vini stepheni
76.6 **Pitcairn Reed-Warbler** . . . . Pi 2
     Acrocephalus vaughani
76.8 **Henderson Island Reed-Warbler** . . Pi 2
     Acrocephalus taiti

81.4

84.2

Northern Marianas Endemics

109 species, including the following 2 endemics:
81.4 **Tinian Monarch** . . . . NMa 2
     Monarcha takatsukasae
84.2 **Golden White-Eye** . . . . NMa 1,3
     Cleptornis marchei

NORTHERN MARIANAS

· 11 URACAS
· 10 MAUG
· 8 ASUNCION
· 7 AGRIGAN
· 6 PAGAN
12 ALAMAGAN
· 9 SARIGAN
· 5 ANATAHAN
1 SAIPAN
2 TINIAN
3 AGIGUAN
4 ROTA
100 km

PALAU

10 KAYANGEL
1 BABELDAOB
2 KOROR
11 MALAKAL
3 NGERUKTABEL
4 MECHERCHAR
5 PELELIIU
6 ANGAUR
7
SONSOROL
8 PULO ANNA
9 MERIR
12 HELEN I.
100 km

Palau Endemics

152 species, including the following 10 endemics:

57.9 **Palau Ground-Dove** . . . . Pa 3–6
       Gallicolumba canifrons
60.1 **Palau Fruit-Dove** . . . . Pa
       Ptilinopus pelewensis
68.7 **Palau Swiftlet** . . . . Pa
       Aerodramus pelewensis
69.8 **Palau Owl** . . . . Pa
       Pyrroglaux podarginus
75.1 **Palau Bush-Warbler** . . . . Pa
       Cettia annae
79.3 **Palau Flycatcher** . . . . Pa 1–5
       Myiagra erythrops
80.1 **Palau Fantail** . . . . Pa 1–5
       Rhipidura lepida
84.1 **Giant White-Eye** . . . . Pa 3,5
       Megazosterops palauensis
84.3 **Dusky White-Eye** . . . . Pa 1–5
       Zosterops finschii
85.5 **Morningbird** . . . . Pa 1–5
       Colluricincla tenebrosa

French Polynesia Endemics

121 species, including the following 24 endemics:

45.8 **Tuamotu Sandpiper** . . . . FrPo 2
   Prosobonia cancellata
57.5 **Polynesian Ground-Dove** . . . FrPo 5–8
   Gallicolumba erythroptera
57.8 **Marquesas Ground-Dove** . . . . . . FrPo 9
   Gallicolumba rubescens
58.4 **Polynesian Imperial-Pigeon** . FrPo 4,11
   Ducula aurorae
58.5 **Marquesas Imperial-Pigeon** FrPo 12,13
   Ducula galeata
59.3 **Grey-Green Fruit-Dove** . FrPo 4,14–19
   Ptilinopus purpuratus
59.4 **Makatea Fruit-Dove** . . . . FrPo 11
   Ptilinopus chalcurus
59.5 **Atoll Fruit-Dove** . . . . . . FrPo 2, excl.11
   Ptilinopus coralensis

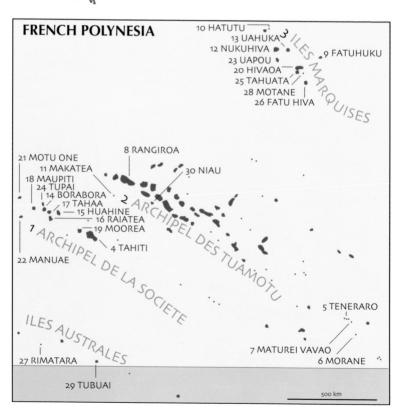

FRENCH POLYNESIA

10 HATUTU
13 UAHUKA
12 NUKUHIVA
23 UAPOU
20 HIVAOA
25 TAHUATA
28 MOTANE
26 FATU HIVA
9 FATUHUKU
ILES MARQUISES

8 RANGIROA
21 MOTU ONE
11 MAKATEA
18 MAUPITI
24 TUPAI
14 BORABORA
17 TAHAA
15 HUAHINE
16 RAIATEA
19 MOOREA
4 TAHITI
22 MANUAE
7
30 NIAU
ARCHIPEL DES TUAMOTU
ARCHIPEL DE LA SOCIETE

ILES AUSTRALES
27 RIMATARA
29 TUBUAI

5 TENERARO
7 MATUREI VAVAO
6 MORANE

500 km

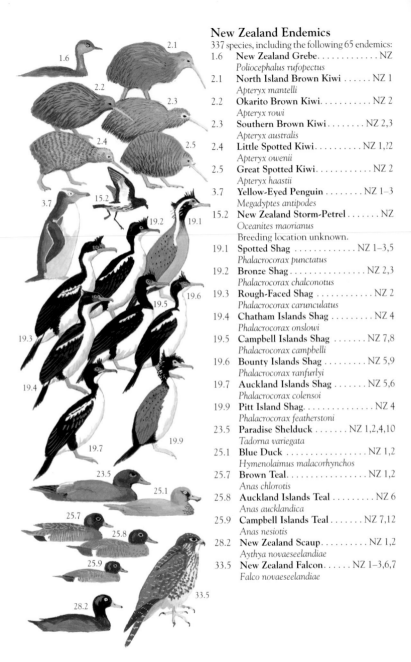

New Zealand Endemics

337 species, including the following 65 endemics:

1.6	**New Zealand Grebe**............NZ	
	Poliocephalus rufopectus	
2.1	**North Island Brown Kiwi**......NZ 1	
	Apteryx mantelli	
2.2	**Okarito Brown Kiwi**..........NZ 2	
	Apteryx rowi	
2.3	**Southern Brown Kiwi**........NZ 2,3	
	Apteryx australis	
2.4	**Little Spotted Kiwi**..........NZ 1,?2	
	Apteryx owenii	
2.5	**Great Spotted Kiwi**...........NZ 2	
	Apteryx haastii	
3.7	**Yellow-Eyed Penguin**........NZ 1–3	
	Megadyptes antipodes	
15.2	**New Zealand Storm-Petrel**.......NZ	
	Oceanites maorianus	
	Breeding location unknown.	
19.1	**Spotted Shag**.............NZ 1–3,5	
	Phalacrocorax punctatus	
19.2	**Bronze Shag**...............NZ 2,3	
	Phalacrocorax chalconotus	
19.3	**Rough-Faced Shag**............NZ 2	
	Phalacrocorax carunculatus	
19.4	**Chatham Islands Shag**.........NZ 4	
	Phalacrocorax onslowi	
19.5	**Campbell Islands Shag**.......NZ 7,8	
	Phalacrocorax campbelli	
19.6	**Bounty Islands Shag**.........NZ 5,9	
	Phalacrocorax ranfurlyi	
19.7	**Auckland Islands Shag**.......NZ 5,6	
	Phalacrocorax colensoi	
19.9	**Pitt Island Shag**..............NZ 4	
	Phalacrocorax featherstoni	
23.5	**Paradise Shelduck**.......NZ 1,2,4,10	
	Tadorna variegata	
25.1	**Blue Duck**.................NZ 1,2	
	Hymenolaimus malacorhynchos	
25.7	**Brown Teal**................NZ 1,2	
	Anas chlorotis	
25.8	**Auckland Islands Teal**.........NZ 6	
	Anas aucklandica	
25.9	**Campbell Islands Teal**......NZ 7,12	
	Anas nesiotis	
28.2	**New Zealand Scaup**..........NZ 1,2	
	Aythya novaeseelandiae	
33.5	**New Zealand Falcon**......NZ 1–3,6,7	
	Falco novaeseelandiae	

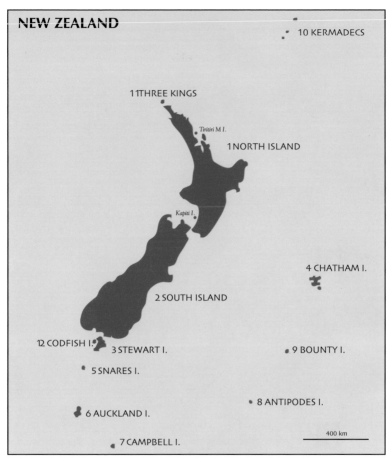

NEW ZEALAND

10 KERMADECS

1 1 THREE KINGS

Tiritiri M I.

1 NORTH ISLAND

Kapiti I.

4 CHATHAM I.

2 SOUTH ISLAND

12 CODFISH I.

3 STEWART I.

5 SNARES I.

9 BOUNTY I.

8 ANTIPODES I.

6 AUCKLAND I.

400 km

7 CAMPBELL I.

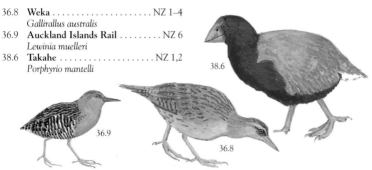

38.6

36.9

36.8

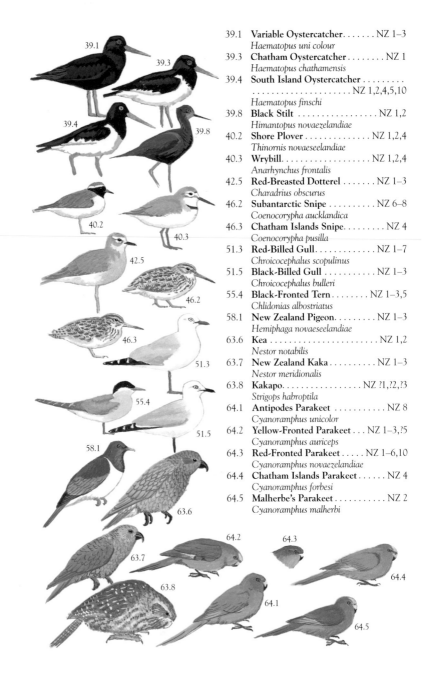

71.1 **Rifleman** . . . . NZ 1–3
Acanthisitta chloris
71.2 **South Island Wren** . . . . NZ 2
Xenicus gilviventris
74.9 **Fernbird** . . . . NZ 1–3
Megalurus punctatus
77.1 **Grey Gerygone** . . . . NZ 1–3,5
Gerygone igata
77.2 **Chatham Island Gerygone** . . . . . . NZ 4
Gerygone albofrontata
77.3 **Yellowhead** . . . . NZ 2
Mohoua ochrocephala
77.4 **Whitehead** . . . . NZ 1
Mohoua albicilla
77.5 **Pipipi** . . . . NZ 2,3
Mohoua novaeseelandiae
80.7 **New Zealand Fantail** . . . . NZ 1–5
Rhipidura fuliginosa
85.1 **Tomtit** . . . . NZ 1–6
Petroica macrocephala
85.3 **Chatham Robin** . . . . NZ 4
Petroica traversi
85.4 **New Zealand Robin** . . . . NZ 2,3
Petroica australis
85.5 **North Island Robin** . . . . NZ 1
Petroica longipes
85.7 **Stitchbird** . . . . NZ 1,11,12
Notiomystis cincta
85.8 **New Zealand Bellbird** . . . . . NZ 1–3,6,7
Anthornis melanura
86.1 **Tui** . . . . NZ 1–6,10
Prosthemadera novaeseelandiae
95.4 **Kokako** . . . . NZ 1,?2,?3
Callaeas cinereus
95.5 **Saddleback** . . . NZ small Is off mainland
Philesturnus carunculatus

The largest number of endemic species is found in New Zealand (65), followed by Hawaii (33), Fiji (28) and French Polynesia (24). New Zealand and Hawaii also have completely endemic families, in New Zealand the kiwis (*Apterygidae*, five species), the New Zealand Wrens (*Acanthissittidae*, two species) and the Wattlebirds (*Callaeidae*, two species) and in Hawaii the family of Hawaiian Creepers (*Drepanididae*, 20 species). All members of the Hawaiian Creepers' family relate back to a finch-like bird, which arrived in Hawaii a long time ago and who's offspring diversified into a large number of species, of which only these 20 remain today. Other bird groups that are represented by many endemics in the area are:

- Rails and Crakes (family *Rallidae*); in total 135 species worldwide, of which 21 are found in the area, including six endemics;
- Ground-Doves (genus *Gallicolumba*); in total 16 species in Philippines, Indonesia and Polynesia, of which six are found in the area, including four endemics;
- Fruit-Doves (genus *Ptilinopus*); in total 51 species in Melanesia, Philippines and Polynesia, of which 15 are found in the area, including 11 endemics;
- Reed-Warblers (genus *Acrocephalus*); in total 37 species worldwide, of which 15 are found in the area, including ten endemics;
- Monarch Flycatchers (family *Monarchidae*); in total 100 species in the tropics, of which 21 are found in the area, including 15 endemics;
- White-eyes (family *Zosteropidae*); in total 96 species in the tropics and subtropics, of which 14 are found in the area, including ten endemics.

The large family of the Parrots *Psittacidae* (worldwide 347 species) is represented by only 28 species, relatively a small amount, of which 16 are endemic and ten are introduced.

Except the 'true' endemics there are also several breeding endemics, especially seabirds who visit countries in the area only to breed but disperse widely after breeding. New Zealand counts 12 breeding endemics, given in the following list:

3.1	**Fiordland Penguin**
	Eudyptes pachyrhynchus
3.2	**Snares Penguin**
	Eudyptes robustus
3.3	**Erect-Crested Penguin**
	Eudyptes sclateri
6.5	**Buller's Albatross**
	Thallasarche bulleri
9.9	**Magenta Petrel**
	Pterodroma magentae
10.1	**Chatham Petrel**
	Pterodroma axillaris
10.8	**Cook's Petrel**
	Pterodroma cookii
11.1	**Pycroft's Petrel**
	Pterodroma pycrofti
12.7	**Westland Petrel**
	Procellaria westlandica
12.9	**Parkinson's Petrel**
	Procellaria parkinsoni
14.2	**Buller's Shearwater**
	Puffinus bulleri
14.5	**Hutton's Shearwater**
	Puffinus huttoni

1 GREBES & LOON

1.1 PIED-BILLED GREBE *Podilymbus podiceps* [Grèbe à bec bigarré] L 34cm. Note heavy bill. Rather vocal. Moults before departure from winter quarters. Wings unmarked in flight. ✳ Winters on sheltered freshwater ponds near vegetation. ♪ Series of nasal, trumpet-like or hollow notes. ☉ Ha.

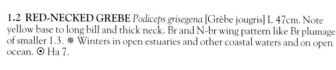

1.2 RED-NECKED GREBE *Podiceps grisegena* [Grèbe jougris] L 47cm. Note yellow base to long bill and thick neck. Br and N-br wing pattern like Br plumage of smaller 1.3. ✳ Winters in open estuaries and other coastal waters and on open ocean. ☉ Ha 7.

1.3 HORNED GREBE *Podiceps auritus* [Grèbe esclavon] L 35cm. From 1.4 by pale tip to bill. Br plumage from 1.4 by colour of neck. Crown of N-br plumage peaks further back than in 1.4. N-br wing pattern like 1.4. ✳ Winters normally on quiet inland waters, sheltered bays and (occasionally) on open ocean. ☉ Ha 7.

1.4 EARED GREBE *Podiceps nigricollis* [Grèbe à cou noir] L 31cm. Differs in N-br plumage from 1.3 by different head silhouette with diffuse streak behind eye. ✳ Winters normally on coastal waters and open ocean. ☉ Ha 2,6,13. V

1.5 HOARY-HEADED GREBE *Podiocephalus poliocephalus* [Grèbe argenté] L 29cm. From 1.3 and 1.4 by range. Rather pale and slim without chestnut tones in plumage (see Br plumage of 1.6). Note clean white cheeks in N-br plumage. ✳ Fresh and saline waters of open wetlands and estuaries. Occasionally on farm ponds. ♪ High, frog-like *puRrreeh* in series of 2–3. ☉ NZ 1,2,5.

1.6 NEW ZEALAND GREBE (or [NZ]Dabchick) *Podiocephalus rufopectus* [Grèbe de Nouvelle-Zélande] L 29cm. Br plumage unmistakable by chestnut neck and silvery streaked head. N-br plumage from similar 1.7 by more elongated jizz and lack of yellow at bill base. ✳ Small, shallow waters such as farm dams and sewage ponds. ♪ Low, raspy *wró'wo'róugh*. ☉ NZ 1,2. E.NZ

1.7 AUSTRALASIAN (or [NZ]Eastern Little) **GREBE**[1] *Tachybaptus novaehollandiae* [Grèbe australasien] L 25cm. Smallest grebe. Yellow bill base and compact build diagnostic. Small water bodies like ponds, dams, quiet streams and swamps. ♪ Very high, twittering *peepeepeep—*. ☉ NZ 1,2.

1.8 GREAT CRESTED GREBE[2] *Podiceps cristatus* [Grèbe huppé] L 50cm. Unmistakable by head pattern. Br and N-br plumage of NZ ssp similar. Imm. without ruff and with short crest. Juv. like all other grebes with striped face and neck sides. ✳ Open lakes. ☉ NZ 1,2. R

1.9 PACIFIC LOON *Gavia pacifica* [Plongeon du Pacifique] L 65cm. Bill normally held horizontally as compared to other loons, but more upturned than most grebes. In N-br plumage with very narrow white eyering and sharp contrast dark-white on neck. Back faintly barred in N-br plumage and with distinct white flash on rearmost part of flanks. ✳ Winters on open ocean and deep bays. ☉ Ha 2,6,17. V

2.1 NORTHERN ISLAND BROWN KIWI *Apteryx mantelli* [Kiwi de Mantell] L 50cm. Note brownish plumage. ✻ Prefers dense forest and shrubland, but occurs also in pine plantations, scrub, second growth and pastures. ♫ High, rising, shrill whistle. ☉ NZ 1. E.NZ

2.2 OKARITO BROWN KIWI *Apteryx rowi* [Kiwi d'Okarito] L 50cm. Greyer than 2.1. ✻ Dense lowland forest. ☉ NZ 2. E.NZ

2.3 SOUTHERN BROWN KIWI[4] *Apteryx australis* [Kiwi austral] L 50cm. From 2.1–2 by range. Sole Brown Kiwi that might come out in daytime. ✻ Wet forest, also in subalpine tussockland. ♫ High, descending, shrieking yell *wee-oh* (♂), that of ♀ deeper and more hoarse. ☉ NZ 2,3. E.NZ.

2.4 LITTLE SPOTTED KIWI *Apteryx owenii* [Kiwi d'Owen] L 40cm. From larger 2.5 mainly by range. ✻ Mainly interior and at margins of forest with dense undergrowth; up to 1,000m. ♫ Long, gradually ascending, high-pitched series of *wreeh* notes. ☉ NZ 1,?2. R E.NZ

2.5 GREAT SPOTTED KIWI *Apteryx haastii* [Kiwi roa] L 55cm. Cf. 2.4. ✻ Mossy beech forest, tussock grassland, coastal scrub. Up to 1,200m, mainly above 700m. ♫ Slow, slightly ascending, mid-high series of *wrraah* notes. ☉ NZ 2. R E.NZ

2.6 KING PENGUIN *Aptenodytes patagonicus* [Manchot royal] L 90cm. From 2.7 by larger yellow base to lower mandible, paler upperparts and narrower black border between dark mantle and white breast. Imm. like Ad. but yellow at bill and throat much paler. ✻ Pelagic, occasionally at islands, rare at mainland. ♫ Thin bleating (as starting of Formula 1 cars). ☉ NZ 2,4–8.

2.7 EMPEROR PENGUIN *Aptenodytes forsteri* [Manchot empereur] L 115cm. Larger than 2.6, which see. Imm. shows white throat, gradually changing with age into black throat like Ad. ✻ Pelagic, rare at islands. ♫ In colony bleating, rising, bouncing *vehvehveh*. ☉ NZ 3.

2.8 GENTOO PENGUIN[5] *Pygoscelis papua* [Manchot papou] L 75cm. Note black eyes, mainly red bill, triangular white patch above eyes and yellow to dull orange feet. ✻ Pelagic, occasionally at islands and mainland. ♫ Low, donkey-like braying. ☉ Off NZ. R

2.9 ADELIE PENGUIN *Pygoscelis adeliae* [Manchot d'Adélie] L 70cm. White eyering diagnostic; eyering lacking in white-chinned Imm. ✻ Pelagic, rare at mainland. ♫ Soft, hoarse trumpeting. ☉ Off NZ.

2.10 CHINSTRAP PENGUIN *Pygoscelis antarcticus* [Manchot à jugulaire] L 75cm. Unmistakable. ✻ Pelagic, rare at islands. ♫ Soft, goose-like squeaks in series of 1–5. ☉ Off NZ.

3 PENGUINS

3.1 FIORLAND (NZCrested) **PENGUIN** *Eudyptes pachyrhynchus* [Gorfou du Fiordland] L 55cm. Note whitish cheek stripes (in most individuals), lack of bare skin at bill base and little black at tips of underflippers. ✳ Breeds in colonies in rainforest or in caves along rocky shores; forages in nearby seas, outside Br season probaby pelagic. ♫ Very high, excited quacking. ⊙ NZ. Absent March–June. R

3.2 SNARES (NZCrested) **PENGUIN** *Eudyptes robustus* [Gorfou des Snares] L 60cm. Note pink bare skin at bill base, robust bill and narrow yellow eyebrow. Black crown feathers not or hardly erectile. Colonies in muddy areas or on rocky flats with shading forest and other vegetation; outside Br season pelagic. ♫ Very high, hoarse quacking. ⊙ NZ. R

3.3 ERECT-CRESTED PENGUIN *Eudyptes sclateri* [Gorfou huppé] L 65cm. From 3.2 by broader, yellow eyebrow, kept erect over eye. ✳ Breeds in rocky areas without vegetation. Outside Br season pelagic. ♫ Very high, nasal, shivering quacking. ⊙ NZ. R

3.4 ROCKHOPPER PENGUIN[6] *Eudyptes chrysocome* [Gorfou sauteur] L 50cm. Rather small posture, small bill, yellow eyebrow spreading fanwise behind eye. Nom. (**a**) differs from ssp *moseleyi* (**b**) by extent of black pattern at underflipper and presence of naked skin at bill base. ✳ Breeds in rocky places, gentle tussock slopes or under vegetation. Outside Br season pelagic. ♫ Low, Mallard-like quacking, sometimes in dancing series. ⊙ NZ. R

3.5 ROYAL PENGUIN *Eudyptes schlegeli* [Gorfou de Schlegel] L 70cm. From other *Eudyptes* penguins by white cheeks. ✳ Colonies mainly on stony ground without vegetation. Pelagic after breeding. ♫ Mid-high, drawn-out braying. ⊙ NZ 1,2,4,5–7 R

3.6 MACARONI PENGUIN *Eudyptes chrysolophus* [Gorfou doré] L 70cm. From 3.5 by all-black head, pink naked skin at bill base and different crest 'style'. Breeds in rocky, flat to steep places, normally without vegetation. Probably pelagic after breeding. ♫ High, Mallard-like quacking, sometimes in short rattles. ⊙ NZ 5,7.

3.7 YELLOW-EYED PENGUIN *Megadyptes antipodes* [Manchot antipode] L 65cm. Unmistakable by yellow eyes and streak from eye to hindcrown. ✳ Breeds in dense vegetation on shores and slopes. Pelagic after breeding. ♫ High-pitched *wrritwrritwrrit*. ⊙ NZ 1–7. R E.NZ

3.8 LITTLE PENGUIN *Eudyptula minor* [Manchot pygmée] L 42cm. Note slaty blue upperparts without distinctive markings to head. Some variation in width of white flipper margins (**a** and **b**). ✳ Small colonies on rocky flats and slopes; also in sand dunes. ♫ Varied: e.g. soft, drawn-out, mumbled croaking or high, rather hoarse screams. ⊙ NZ 1–4.

3.9 MAGELLANIC PENGUIN *Spheniscus magellanicus* [Manchot de Magellan] L 70cm. Unmistakable by head pattern and double breast-band. ✳ Vagrant from S America to bays of NZ and Australia. ♫ Braying sneezes. ⊙ NZ 1.

4 ALBATROSSES

4.1 WANDERING ALBATROSS[7] *Diomedea exulans* [Albatros hurleur]
W 303cm. Separation from 4.2 only possible at close range when lack of black line along cutting edge of upper mandible visible. Field identification to ssp and age on base of plumage features in most cases not possible. Passes through about 7 stages (stages (St) 1–4 and 7 shown, stages 5 and 6 not shown) before attaining full Ad. plumage. This process, that can take up to 16 years, can stop in any stage so that for instance some ♀♀ can retain parts (e.g. black tail feathers) or the complete Juv. plumage (stage 1) during their whole life. ♂♂ tend to become whiter than ♀♀. Shown Nom. (**a**), ssp *gibsoni* (**b**) and ssp *antipodensis* (**c**).
✻ Open ocean. ♫ Varied, e.g. very low purring, bill rattling, nasal gackling, etc.
☉ Breeds NZ 6–8.

4.2 ROYAL ALBATROSS[8] *Diomedea epomophora* [Albatros royal] W 328cm.
From 4.1 by black cutting edge of bill. In all but Juv. stage the tail is white (unlike 4.1), so a large albatross with white wing patch combined with all-white tail is always this species. No dark brown Juv. stage and less Imm. stages than 4.1, especially in ssp *sanfordi* (**a**, Nothern Royal Albatross) where Juvs moult direct into Ad. plumage. Note difference between 4.2**a** with solid black wings and Nom. (**b**, Southern Royal Albatross); in the latter the wings of Ad. are mainly white. Breeds on flat or gently sloping ground. Nests sheltered by rocks or low vegetation but needs gently sloping, exposed sites for take-off and landing. Otherwise open ocean. ♫ Low, rhythmic or puffed-out grunts and mumblings. Also bill rattling. ☉ Breeds NZ 4,6,7.

4.3 KELP (or [NZ]Southern Black-backed) **GULL** *Larus dominicanus* [Goéland dominicain] L 60cm. Frontal view for comparison of size with 4.1–2. See also 52.1.

4.4 BLACK-BROWED ALBATROSS *Thalassarche melanophris* [Albatros à sourcils noirs] W 230cm. Frontal view for comparison of size with 4.1–2. See also 6.3.

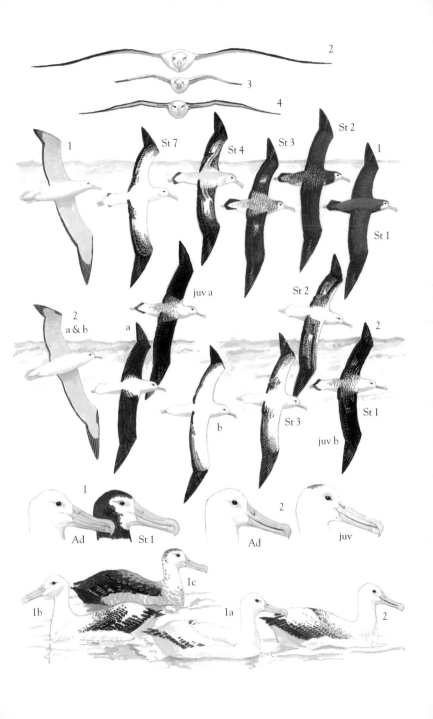

5 ALBATROSSES

5.1 BLACK-FOOTED ALBATROSS *Phoebastria nigripes* [Albatros à pieds noirs] W 203cm. Note white at bill base and in Ad. (showing more white at face sides than Juv.) white under- and uppertail coverts. Ad. can look very bleached-out overall. Bill and feet dusky pink. Cf. Juv. 5.3 (with bright pink bill and paler underside of flight feathers) and Juv. of much larger 4.1 (with white face mask). ✳ Offshore and open ocean. ♫ Duck-like qaucking, thin, nasal mumbling and high, smooth squeaks. Also bill clapping. ☉ Breeds NW Ha, Ma. Wanders S to Equator. V NZ

5.2 LAYSAN ALBATROSS *Phoebastria immutabilis* [Albatros de Laysan] W 199cm. Note dark patch around eyes. Underwings can vary like shown (**a** and **b**, not age related). Note range. ✳ Open ocean. ♫ Bleating, shivering calls and bill snapping. ☉ Breeds NW Ha. Wanders mainly N.

5.3 SHORT-TAILED ALBATROSS *Phoebastria albatrus* [Albatros à queue courte] W 221cm. Yellowish head of Ad. diagnostic. More extensive than in 4.1 and 4.2, that often show yellow smear between cheeks and nape. Underside of flight feathers in Juv. rather pale. Note characteristic pattern of upperwings like shown in Imm. and Ad. ✳ Open ocean. ☉ Off Ha. R

6 ALBATROSSES

6.1 GREY-HEADED ALBATROSS *Thalassarche chrysostoma* [Albatros à tête grise] W 192cm. From 6.2 by greyer head, more black to leading edge of underwing and yellow lower ridge to bill. Rather similar to 6.4, but note that yellow starts narrowly at upper ridge of bill. Juv. with dark underwings like Juv. 6.3, but differs by white, not grey head. ✱ Offshore and open ocean. ♫ Mid-high, nasal bleating. ☉ Breeds NZ 7. R

6.2 YELLOW-NOSED ALBATROSS[9] *Thalassarche chlororhynchos* [Albatros à nez jaune] W 200cm. Shown are Nom. (**a**, with grey head) and ssp *bassi/carteri* (**b**, with white head). From 6.4 by lack of black mark in axillaries. Lack of yellow line along lower mandible in Ad. diagnostic. Note narrow black leading edge of underwings. ✱ Inshore and offshore. ♫ High, scratchy bleating or raspy *wuc-wuc-wuc*. ☉ Off NZ.

6.3 BLACK-BROWED ALBATROSS[10] *Thalassarche melanophris* [Albatros à sourcils noirs] W 230cm. Shown are dark-eyed Nom. (**a**) and pale-eyed ssp *impavida* (**b**). **b** might show slightly more extensive black than **a** to underwings (not shown). Yellow bill and lack of grey to head diagnostic in Ad. Juv. with black bill and dark underwings shows complete or partial grey breast-band. ✱ Breeds on cliff ridges and ledges. Offshore and pelagic. ☉ Breeds NZ 5,7,8.

6.4 SHY (or [NZ]White-capped) **ALBATROSS**[11] *Thalassarche cauta* [Albatros à cape blanche] W 234cm. 4 ssps in the area: Nom. ([NZ]Tasmanian Albatross not shown; resembles **b**), *steadi* (**a**, [NZ]New Zealand White-capped Albatross), *salvini* (**b**, [NZ]Salvin's Albatross) and *eremita* (**c** [NZ]Chatham Island Albatross). **a**, **b** and **c** breed in the area. Black diagnostic mark in axillaries, where black leading edge of wings meets body. Colour of bill and saturation of grey or brown to head differ among ssps, but all ssps show narrow black leading edge to underwings. ✱ Breeds on level places in rocky, broken areas. Offshore. ♫ High, drawn-out, slightly shivering, lowing or toneless, nasal twittering and braying. ☉ Breeds NZ 4–6,8. R

6.5 BULLER'S ALBATROSS *Thalassarche bulleri*[12] [Albatros de Buller] W 209cm. Two ssps: Nom. (**a**, Northern Buller's Albatross): note broad yellow base to yellow line along culmen; and ssp *platei* (**b**, [NZ]Southern Buller's Albatross, with a greyer forehead). **b** differs from **a** mainly in timing of breeding (**a** breeds 2–3 months later than **b**). Bill of **b** is supposed to be slightly broader, black eyebrow longer between eye and base of bill and forehead and crown silvery grey, not silvery white like **a**, but all these features may vary. From Ad. 6.1 also by clear-cut black leading edge to underwings. ✱ Breeds on slopes, ridges and tops of rocky slopes with sparse, low vegetation. Inshore, offshore and open ocean. ♫ Braying notes and drawn-out lowing. ☉ **a** breeds NZ 4,5,12; **b** breeds NZ 4,11. R

7.1 SOOTY ALBATROSS *Phoebetria fusca* [Albatros brun] W 203cm. Overall sooty brown, slightly darker on head. In worn plumage (**b**) paler on back, especially in neck, and then slightly similar to 7.2. Note that line along lower mandible is pale yellow. This line in Juv. very narrow and hardly visible. ✳ Vagrant over open sea. ☉ Off NZ. R

7.2 LIGHT-MANTLED (NZSooty) **ALBATROSS** *Phoebetria palpebrata* [Albatros fuligineux] W 200cm. Body and underwings paler-coloured than 7.1, especially on mantle. Line along lower mandible is white and narrower than that of 7.1. Juv. scalloped above and line along lower mandible hardly visible or absent. ✳ Rocky islands with vegetated cliffs and steep slopes. ♪ Drawn-out wails. ☉ Breeds NZ 6–8. R

7.3 HALL'S (or NZNorthern) **GIANT PETREL** *Macronectes halli* [Pétrel de Hall] W 190cm. Very similar to 7.4; main difference is colour of bill tip which in this species is red-brown. Eyes normally pale (sometimes brown), becoming even paler with age. No white plumage morph like 7.4. Imm. dark brown like Imm. 7.4, but with reddish bill tip. ✳ Coastal areas with some shelter of sparse vegetation, rocks, banks. Inshore, offshore and open ocean. ♪ Low, nasal grumbles and croaks. ☉ Breeds NZ 4,6–8. R

7.4 ANTARCTIC (or NZSouthern) **GIANT PETREL** *Macronectes giganteus* [Pétrel géant] W 195cm. From 7.3 mainly by pea-green bill tip. White morph (**a**, lacking in 7.3) is uncommon. Eyes normally brown but can be pale grey. ✳ Inshore and offshore. ♪ Low, nasal grumbles and croaks. ☉ NZ seas. R

8 PRIONS, FULMARS & PETRELS

Note that most Prions show distinct M on wings, but cf. 9.3–4, 9.7–8, 10.1–2, 10.5–8, 11.1, 11.4–5, 12.3 and 14.2.

8.1 BROAD-BILLED PRION *Pachyptila vittata* [Prion de Forster] W 62cm. Seemingly large-headed. Bill larger than the other 5 prions and coloured black (not grey like 8.2). Note steep forehead. ✳ Breeds in burrows or in caves and crevices in cliffs. Offshore and open ocean. ♪ Short, very low, rhythmic croaking. ☉ Breeds NZ 4,5.

8.2 SALVIN'S PRION *Pachyptila salvini* [Prion de Salvin] W 57cm. Separation from 8.1 and 8.3 probably impossible at sea. In the hand shows comb-like lamellae at base of closed bill. Bill grey. ✳ Inshore and offshore. ♪ Hoarse, rather excited croaking at different pitch. ☉ NZ seas.

8.3 ANTARCTIC PRION *Pachyptila desolata* [Prion de la Désolation] W 63cm. Smaller and bluer bill than 8.1 and – though collar more distinct – not safely separable from 8.2. ✳ Breeds in burrows and rock crevices; feeds offshore and in open ocean. ♪ Low, hoarse and falsetto croaking (not unlike statics on radio). ☉ Breeds NZ 6,8.

8.4 SLENDER- (or NZThin-)**BILLED PRION** *Pachyptila belcheri* [Prion de Belcher] W 56cm. The M on wings less distinct than 8.1–8.3, especially in worn plumage (**a**). Note small, narrow bill and extensive, distinct white eyebrow, giving pale-faced expression. ✳ Inshore, offshore and open ocean. ☉ NZ seas.

8.5 FAIRY PRION *Pachyptila turtur* [Prion colombe] W 56cm. Note indistinct facial pattern and narrow M on wings. At sea not safely separable from 8.6. ✳ Breeds in burrows or in rock crevices and caves. Feeds offshore and in open ocean. ♪ High, excited croaking. ☉ Breeds NZ 4,5,8.

8.6 FULMAR PRION[13] *Pachyptila crassirostris* [Prion à bec épais] W 60cm. Difficult to separate at sea from smaller-billed 8.5 with similar pale face pattern; best distinction said to be its 'looping' flight. ✳ Breeds in crevices and in tunnels in cliffs. At sea, inshore and offshore. ♪ Low, hoarse, at times bouncing croaking. ☉ Breeds NZ 4–6,9.

8.7 ANTARCTIC PETREL *Thalassoica antarctica* [Pétrel antarctique] W 105cm. Note characteristic pattern of upperparts with contrasting white secondaries and tail. ✳ Inshore and offshore. ♪ Low, hoarse croaks. ☉ Breeds NZ 8. R

8.8 NORTHERN FULMAR *Fulmarus glacialis* [Fulmar boréal] W 107cm. Pale morph (**b**) more blotchy, less solid pearl-grey above and its dark morph (**b**, only morph yet seen near Hawaii) is darker than the Atlantic ssp *glacialis* (which does not occur in the area). ✳ Offshore. ☉ Off Ha. R

8.9 SOUTHERN (or NZAntarctic) **FULMAR** *Fulmarus glacialoides* [Fulmar argenté] W 118cm. Note diagnostic contrasting wing pattern with white flash on outer wings more contrastingly patterned above than 8.4. From gulls by gliding flight on stiff wings. ✳ Offshore and open ocean. ♪ High, hoarse, subdued shrieks. ☉ Breeds NZ 8.

9 PETRELS

9.1 TAHITI PETREL *Pterodroma rostrata* [Pétrel de Tahiti] W 84cm. Very similar to 10.3, which see. Note the pale stripe down centre of underwings, sharp demarcation on chest and rather pale uppertail coverts. Flies with straight wings held perpendicular to body, alternating deep wingbeats with banking glides and arcs. ✳ Breeds in burrows on mountain slopes with thick forest cover. Forages at open sea. ♫ Nasal, drawn-out shrieks at varying pitch. ☉ Breeds FrPo 3,4,19, ?Co 1. R

9.2 FIJI PETREL *Pterodroma macgillivrayi* [Pétrel des Fidji] W 68cm. Characterised by small size and uniform brown plumage. From 12.5 by lack of paler wingbar. ✳ Breeding grounds unknown; presumed to forage offshore. ☉ Fi 10. R E.Fi

9.3 GOULD'S PETREL[14] *Pterodroma leucoptera* [Pétrel de Gould] W 70cm. Note small size, black cap, narrow underwing markings and dark M across wings. ✳ Offshore and open seas. ♫ Very high, sharp, nasal shrieks *tfeet-tfeet—*. ☉ A.Sa 2. Fi, Co. R (see also 11.4).

9.4 BLACK-WINGED PETREL *Pterodroma nigripennis* [Pétrel à ailes noires] W 67cm. From 9.3 by paler cap and darker, broader underwing markings. Offshore and open seas. Digs nest burrows under scrub and tussocks or breeds in rock crevices. ♫ High, hollow, mourning *twoot* notes in series of 3–20. Also high, nasal and falsetto shrieks. ☉ Breeds NZ 4,10,11, ?FrPo 30.

9.5 PROVIDENCE PETREL *Pterodroma solandri* [Pétrel de Solander] W 100cm. Head contrastingly darker. Note white markings at bill base and pale base of primaries on underwings. Very similar to 9.6, 10.4 (dark morph), 11.2 and 11.3, which see. ✳ Offshore and open seas. Nests in burrows or rock crevices, often within forest. Breeds also on sparsely vegetated cliffs. ♫ High, upslurred *eeuuh-eeuuh—*. ☉ Disperses N after breeding Lord Howe I. R

9.6 GREAT-WINGED PETREL[15] *Pterodroma macroptera* [Pétrel noir] W 99cm. Nom. (**a**, with little white at bill base) shown and ssp *gouldi* (**b**, [NZ]Grey-faced Petrel, with more white in face). From 9.5 by more uniform plumage and little or no pale flash on outer underwings. ✳ Offshore and open ocean. Nests in burrows or crevices under clumps of vegetation. ☉ Breeds islands and mainland N NZ 1.

9.7 SOFT-PLUMAGED PETREL *Pterodroma mollis* [Pétrel soyeux] W 89cm. Polymorph, pale (**a**) and rare dark (**b**) morphs shown. Note that collar is closed across breast and that more or less distinct pattern of underwings (often appearing uniform dark) resembles pattern of upperwings. ✳ Offshore and open ocean. Breeds in burrows under dense cover. ♫ Very high, drawn-out, wailing shrieks. ☉ Breeds NZ 8.

9.8 WHITE-HEADED PETREL *Pterodroma lessonii* [Pétrel de Lesson] W 109cm. Note white head and pale grey rump and tail, contrasting with darker wings and mantle. ✳ Offshore and open ocean. Breeds in burrows under tussocks, ferns, low shrub, often in open grassland. ♫ Low croaks and series sounding as if produced by a dry bicycle pump. ☉ Breeds NZ 6,8.

9.9 MAGENTA (or [NZ]Chatham I. Taiko) **PETREL** *Pterodroma magentae* [Pétrel de Magenta] W 102cm. From 9.1 by different underwing pattern, from 10.3 mainly by range. ✳ Offshore. Breeds in dense forest. ☉ Breeds NZ 4. R

10 PETRELS

10.1 CHATHAM PETREL *Pterodroma axillaris* [Pétrel des Chatham] W 67cm. Diagnostic underwing pattern, in which carpal joint is connected by bar to rear edge where touching body. ❋ Offshore and open ocean. Breeds in burrows in soft clay under vegetation clumps and tussocks. ♫ Very high, rapid, staccato *tseetseetsee* in flight. ☉ Breeds NZ 4. R

10.2 MOTTLED PETREL *Pterodroma inexpectata* [Pétrel maculé] W 85cm. Underparts with diagnostic broad bar across underwings and brownish belly. ❋ Nests in burrows in rocky ground and tussock grassland. ♫ High, thin shrieks in short, slightly lowered series. ☉ Breeds NZ 3,5. R

10.3 PHOENIX PETREL *Pterodroma alba* [Pétrel à poitrine blanche] W 83cm. Very similar to 9.1, 9.9 and 11.3, but underwings solid brown except narrow white line along leading edge. ❋ Offshore and open ocean. Breeds not in burrows, but on the ground at sheltered sites under trees or bushes. ♫ High, sharp bickering, down-slurred at the end. ☉ Breeds Ki, FrPo 3, Pi. R

10.4 KERMADEC PETREL *Pterodroma neglecta* [Pétrel des Kermadec] W 92cm. Polymorph, dark (**a**) morph and pale (**b**) morphs shown; most common is intermediate morph (not shown). All morphs generally difficult to separate from similar petrels (e.g. 9.5, 11.2 and 11.3) but white primary shafts on upperwings (hard to see) diagnostic. ❋ Nests on cliffs or slopes with little vegetation. ♫ Long, drawn-out, mourning shrieks and bleating, changing in pitch. ☉ Breeds NZ 10. R

10.5 JUAN FERNANDEZ PETREL *Pterodroma externa* [Pétrel de Juan Fernandez] W 97cm. Note large size and narrow black bar to underwings. Often shows horseshoe-like white base to upper tail. ❋ Breeds outside the area. ♫ Short series of mewing and trumpet-like notes. Also drawn-out hooting. ☉ Breeds Juan Fernandez Is, Chile, from where disperses north. R

10.6 WHITE-NECKED (or ᴺᶻWhite-naped) **PETREL** *Pterodroma cervicalis* [Pétrel à col blanc] W 100cm. White neck diagnostic but beware of individuals of 10.5 with pale grey neck. ❋ Offshore and open seas. May nest in burrows (Kermadecs). ☉ Breeds NZ 10. R

10.7 STEJNEGER'S PETREL *Pterodroma longirostris* [Pétrel de Stejneger] W 60cm. Note narrow black bar to underwings, similar to 10.8, but shows darker upperparts. ❋ Offshore and open sea. Nests in burrows under ferns. ☉ Breeds Juan Fernandez Is, Chile, from where disperses north. R

10.8 COOK'S PETREL *Pterodroma cookii* [Pétrel de Cook] W 65cm. Cf. 10.7. Upper parts paler grey than other small petrels. ❋ Breeds in burrows near ridges or steep slopes in thick forest. ♫ Mid-high, nasal chatters. ☉ Breeds NZ 12 and other small islands off NZ. R

11 PETRELS

11.1 PYCROFT'S PETREL *Pterodroma pycrofti* [Pétrel de Pycroft] W 53cm. Note small size and narrow black line from wrist to body. In general darker than 10.8 and with less extensive dark patch around eye than 10.7. ✳ Offshore. Nests in burrows in forest. ♪ Very high *peepeep-puh* (last part much lower). ☉ Main colonies at islands off N NZ 1. R

11.2 MURPHY'S PETREL *Pterodroma ultima* [Pétrel de Murphy] W 97cm. From 10.4 (dark morph) by indistinct M pattern on upperwings and lack of white primary shafts. Offshore and open seas. Breeds openly on the ground, usually next to some wind protection. ♪ High barking and low, hollow, drawn-out wailing. ☉ Breeds FrPo, Co, Pi.

11.3 HERALD PETREL *Pterodroma arminjoniana* [Pétrel hérault] W 95cm. Polymorph, most common are intermediates (not shown) between dark (**a**) and pale (**b**) morphs. Note underwing pattern. ✳ Offshore and open seas. Surface breeder at sheltered sites. ♪ Drawn-out whinnying and bickering. ☉ Breeds Ton, Co 1, FrPo, Pi.

11.4 GOULD'S PETREL *Pterodroma leucoptera* [Pétrel de Gould] W 70cm. Note blackish cap, indistinct M pattern on upperwings and broad black mark along leading edge of underwings. ✳ Offshore and open seas. Nests in gullies and on slopes with palm cover. ♪ Very high, sharp, nasal shrieks *tfeet-tfeet—*. ☉ A.Sa 2, Fi, Co. R (see also 9.3).

11.5 BONIN PETREL *Pterodroma hypoleuca* [Pétrel des Bonin] W 67cm. Underwing pattern diagnostic. ✳ Offshore and open seas. Burrow breeder in sandy soils with vegetation. ♪ Low, very soft, rattling croaks. ☉ Breeds Ha 13–18, NMa.

11.6 HAWAIIAN PETREL *Pterodroma sandwichensis* [Pétrel des Hawaï] W 91cm. Note characteristic dark cap and lack of M pattern on upperwings. ✳ Offshore and open seas. Breeds in burrows or crevices under vegetation high on mountain slopes. ♪ High, hollow, three-syllabled hoots and mid-high, slightly squeaking or croaking calls. ☉ Breeds Ha 1,2,4,?5,?7. R E.Ha

11.7 HENDERSON PETREL *Pterodroma atrata* [Pétrel de Henderson] W 90cm. From dark 10.4 by lack of white primary shafts. ✳ Surface nester in dense forest. ♪ Undulating shivering twitter. ☉ Breeds Pi.

12 PETRELS

12.1 CAPE PETREL[16] *Daption capense* [Damier du Cap] W 85cm. Unmistakable by upperparts pattern. ✳ Offshore and open seas. Forages inshore near Br colonies. Nests on level ground, where preferring overhanging rock. ♫ E.g. mid-high barks and trills. ⊙ Breeds NZ 4,6–9.

12.2 SNOW PETREL[17] *Pagodroma nivea* [Pétrel des neiges] W 85cm. No similar bird in the area. ✳ Offshore. ♫ Toneless croaks and bickering. ⊙ Only seen far S from NZ.

12.3 BLUE PETREL *Halobaena caerulea* [Prion bleu] W 65cm. From *Pterodroma* petrels by lack of dark bar to underwings, from prions by white tail tip. ✳ Offshore and open ocean. Breeds in burrows in coastal slopes with tussocks. ♫ Mid-high, cooing notes. ⊙ At sea around NZ. R

12.4 BULWER'S PETREL *Bulweria bulwerii* [Pétrel de Bulwer] W 71cm. Entirely dark brown with paler bar across upperwings. Note long-winged and - tailed jizz. ✳ Inshore near Br colonies. Breeds in crevices and burrows. ♫ Low, rhythmic, sustained barking. ⊙ Breeds Ha, Ma, FrPo 3.

12.5 JOUANIN'S PETREL *Bulweria fallax* [Pétrel de Jouanin] W 79cm. All dark with indistinct pale bar across upperwings. Note somewhat oversized bill. ✳ Open ocean. ⊙ Beach wreck Ha 14,17.

12.6 GREY PETREL *Procellaria cinerea* [Puffin gris] W 123cm. Note dark underwings, white underparts and capped appearance. ✳ Offshore and open seas. Burrow breeder in steep vegetated slopes. ♫ Series of rapid, goose and ducklike croaking. ⊙ Breeds NZ 7,8. R

12.7 WESTLAND PETREL *Procellaria westlandica* [Puffin du Westland] W 137cm. From 12.8 by black bill tip and white chin (not always present but if so difficult to see). Larger than 12.9. ✳ Offshore. Burrow breeder in forested hills. ♫ Hoarse croaking, changing to long chatter. ⊙ Breeds NZ 2 (W coast). R

12.8 WHITE-CHINNED PETREL *Procellaria aequinoctialis* [Puffin à menton blanc] W 140cm. Small white chin mark diagnostic, but this feature may be missing. ✳ Offshore and open seas. Inshore near Br colonies. Nests in burrows in open terrain. ♫ E.g. bill rattle, changing to and ending in gull-like shriek. ⊙ Breeds NZ 6–8. R

12.9 PARKINSON'S (or [NZ]Black) **PETREL** *Procellaria parkinsoni* [Puffin de Parkinson] W 115cm. Bill shows dark tip. Note blackish, not pink feet. ✳ Offshore and open seas. Breeds in burrows in forest. ♫ High, level, rapid, sharp chatter. ⊙ Breeds small islands off N NZ 1. R

13 PETREL & SHEARWATERS

13.1 KERGUELEN PETREL *Aphrodroma brevirostris* [Pétrel de Kerguelen] W 81cm. Note fairly thickset jizz. Distinctive underwing pattern with silvery underside of primaries and pale leading edge between body and wrist. Arcs higher up in flight than other petrels. ❋ Offshore and open ocean. Breeds in burrows, excavated in wet soil of marshes or lava ridges. ♫ High, hoarse, trumpetlike shrieks or very high piercing screams. ☉ Seas around NZ.

13.2 CORY'S SHEARWATER *Calonectris diomedea* [Puffin cendré] W 112cm. Rather pale upperparts, tail darkest. Bill yellowish with dark tip. Like 13.3 often with down-bended wings. ❋ Inshore, offshore and open ocean. Breeds near S Europe and W Africa. ♫ Series of low shrieks, each one inhaled and rising. ☉ Beach wreck NZ 1. R

13.3 STREAKED SHEARWATER *Calonectris leucomelas* [Puffin leucomèle] W 122cm. Note lack of darker eyebrow, pinkish bill and pattern of outer underwings. May show whitish crescent at base of tail. ❋ Inshore and offshore. Burrow breeder in forest. Climbs trees for takeoff. ☉ Breeds Japan, Taiwan. R

13.4 CHRISTMAS ^{NZ}Island **SHEARWATER** *Puffinus nativitatis* [Puffin de la Nativité] W 76cm. Uniform sooty black, differing from other shearwaters within range by smaller size and darker underwings. ❋ Offshore and open seas. Surface breeder in hollows and crevices or under bushes and in other sheltered sites. ♫ Low nasal croaking *oOoh*. ☉ Breeds e.g. Ha, Ma, Pi. R

13.5 PINK-FOOTED SHEARWATER *Puffinus creatopus* [Puffin à pieds roses] W 109cm. Overall dull-coloured with pink feet and bill. From 13.2–3 by underwing pattern. Note wings held straight out. ❋ Offhore and open seas. Nests in burrows in forested or open, hilly terrain. ♫ Strange, lamenting muttering. ☉ Ki 3, Ma, NZ, Ha. Breeds islands off Chile, from where disperses N. R

13.6 SOOTY SHEARWATER *Puffinus griseus* [Puffin fuligineux] W 102cm. Note extensive white areas on underwings. ❋ Offshore. Nests in burrows under tussocks or dense scrub, sometimes in forest. ♫ Low, pumping muttering. ☉ Breeds many islands and mainland headlands of NZ.

13.7 SHORT-TAILED SHEARWATER *Puffinus tenuirostris* [Puffin à bec grêle] W 98cm. Like 14.6 with white areas to underwings, but these mainly restricted to outer wings. Note protruding feet. Shorter bill than 13.6. ❋ Offshore. Breeds in burrows in open, grassed or scrubby terrain. ♫ Hoarse shrieking and muttering. ☉ Breeds Australia.

13.8 TOWNSEND'S SHEARWATER *Puffinus auricularis* [Puffin de Townsend] W 79cm. Distinctive white flank mark. Probably not in range of 13.9, which is less black on upperparts. ❋ Offshore and open ocean. Breeds in burrows in areas with bush, grass and bracken, often at forest edge. ☉ Seen up to 1,300km SW of Ha.

13.9 AUDUBON'S SHEARWATER *Puffinus lherminieri* [Puffin d'Audubon] W 71cm. Note white markings around eyes and small size. ❋ Mainly offshore. Breeds in hollows, crevices or on cliffs. ♫ High, shrill, drawn-out rattling shrieks. ☉ Breeds e.g. NMa, Pa, Sa, Ton, FrPo.

14 SHEARWATERS

14.1 FLESH-FOOTED SHEARWATER *Puffinus carneipes* [Puffin à pieds pâles] W 103cm. Note pinkish bill with dark tip, pink legs and silvery reflections to underside of flight feathers. ❋ Offshore and open ocean. Nests in burrows in grassland or forest. ♫ High, slow, lamenting barks. ☉ Breeds mainly S Australia and islands off NZ 1,2.

14.2 BULLER'S SHEARWATER *Puffinus bulleri* [Puffin de Buller] W 98cm. Distinctive pattern of upperparts and black tail tip diagnostic. ❋ Offshore and open seas. Nests in burrows or crevices preferring densely forested slopes. ♫ Muttered, nasal cries. ☉ Breeds islands off NZ 1. R

14.3 WEDGE-TAILED SHEARWATER *Puffinus pacificus* [Puffin fouquet] W 101cm. Polymorph, dark morphs (a) most common S of equator. Wedge-shaped tail (only visible when spread) diagnostic. Note uniform dark underwings of dark morph. Dark morph from 14.1 by darker bill and more slender jizz, pale morph (b) from 13.5 by cleaner underwings. ❋ Offshore and open seas. Nests in burrows or on ground in forest or grassland. ♫ Drawn-out, moaning, nasal calls. ☉ Breeds Ha, NMa, Ma, Ki, Fi, Ton, Sa, FrPo, Pi. R

14.4 GREAT(ER) SHEARWATER *Puffinus gravis* [Puffin majeur] W 109cm. Unmistakable by clean-cut cap, white neck and white crescent at tail base. ❋ Offshore and open seas. Rarely inshore. ☉ NZ 2,4.

14.5 HUTTON'S SHEARWATER *Puffinus huttoni* [Puffin de Hutton] W 81cm. Very similar to 14.5–8, all 4 species with black at rump narrowed just after wings. This species (14.5) shows darkest underwings; cf. 14.6. ❋ Offshore. Nests in burrows in mountain slopes with tussocks and scrub. ♫ Excited, exhausted sounding panting. ☉ Breeds NZ 2 (Kaikoura). R

14.6 FLUTTERING SHEARWATER *Puffinus gavia* [Puffin volage] W 76cm. From 14.5 by less dark underwings but note dark markings in axillaries. ❋ Inshore, also offshore. Breeds in burrows in forest or grassland. ♫ Excited, rather rapid chattering. ☉ Breeds small islands NZ 1,2 (N end).

14.7 MANX SHEARWATER[18] *Puffinus puffinus* [Puffin des Anglais] W 82cm. Note pale crescent behind black ear patch and indistinct black margins on underwings. ❋ Offshore. From Atlantic Ocean. ♫ Bleating calls at varying pitch, interspersed with short magpie-like notes. ☉ NZ 1.

14.8 LITTLE SHEARWATER[19] *Puffinus assimilis* [Puffin semblable] W 62cm. Shown are Nom. (a, with dark smear below eye) and ssp *elegans* (b, no black below eyes); ssps *kermadecensis* and *haurakiensis* (both not shown) like b, but *haurakiensis* may show faint smear below eyes. Underwings white without markings except darker tip and trailing edge. Flaps, rarely glides. ❋ Inshore, also offshore. Breeds in burrows or on ground, protected by scrub or rocks. ♫ Low, hoarse panting and indignant muttering. ☉ NZ 1,4,8,10.

15 STORM-PETRELS

15.1 POLYNESIAN (or White-throated) **STORM-PETREL** *Nesofregetta fuliginosa* [Océanite à gorge blanche] W 54cm. Polymorphic, pale morph most common thr., but in Samoa only dark morph seen; all morphs on the Phoenix Is. Largest storm-petrel. Offshore and open ocean. Nests on ground between boulders or in burrows under scrub or in crevices. ♫ High, well-separated *pew-pew*, 2nd note high and staccato. ☉ Breeds e.g. Fi, Ki, FrPo. R

15.2 NEW ZEALAND STORM-PETREL *Pealeornis maoriana* [Océanite de Nouvelle-Zélande] W 44cm. Rediscovered in 2003. From other storm-petrels by streaked underparts. ❀ Offshore. ☉ E.NZ; breeding location unknown.

15.3 GREY-BACKED STORM-PETREL *Garrodia nereis* [Océanite néréide] W 39cm. Unmistakable by grey-and-dark brown pattern of upperparts. ❀ Offshore. Nests in dense ground vegetation. ♫ Series of level, well-separated, hoarse, toneless shrieks. ☉ Breeds NZ 4,6,7,8.

15.4 WILSON'S STORM-PETREL *Oceanites oceanicus* [Océanite de Wilson] W 40cm. Note predominantly dark plumage, white rump, faint wingbar and slightly paler patch on underwings. ❀ Inshore and offshore. Nests in crevices or between boulders. ♫ Series of short, very high to toneless, rasping barks. ☉ Breeds in Antarctica.

15.5 WHITE-FACED STORM-PETREL[20] *Pelagodroma marina* [Océanite frégate] W 42cm. Unmistakable by white face and long legs. ❀ Offshore and open ocean. Breeds in burrows under vegetation. ♫ Series of mid-high, short barks. ☉ Breeds many islands off NZ. *Note:* Shown is ssp *dulciae* [NZ] Australian White-faced Storm Petrel (V off NW NI NZ); not shown is ssp *maoriana* [NZ] New Zealand White-faced Storm Petrel (breeds Is off NZ; with forked tail and often complete breast band) and ssp *albiclunis* [NZ] Kermadec White-faced Storm Petrel (supposed to have white rump, but such birds might also be fresh-plumaged other races).

15.6 WHITE-BELLIED STORM-PETREL *Fregetta grallaria* [Océanite à ventre blanc] W 46cm. Very variable, from white belly and underwing coverts (**a**) via intermediate morph (**b**) to all-dark morph (**c**). Cf. 15.4 which differs from **c** by white crescent across rump and paler underwing patch. 15.6**b** is very similar to 15.7 but rump less white. ❀ Offshore and open ocean. Breeds in earth holes and crevices. ♫ High, thin notes like *feeeh* or *tjew*. ☉ Breeds NZ 10. R

15.7 BLACK-BELLIED STORM-PETREL *Fregetta tropica* [Océanite à ventre noir] W 46cm. With varying underparts, most with (**a**), some without black to belly (**b**), ressembling 15.6**b**, but rump crescent always pure white. ❀ Offshore and open ocean. Breeds in small holes and crevices. ♫ Very high, thin *feet* and short, oystercatcher-like *twit twittwit -*. ☉ Breeds e.g. NZ 6,8.

16 STORM-PETRELS & DIVING-PETRELS

16.1 SWINHOE'S STORM-PETREL *Oceanodroma monorhis* [Océanite de Swinhoe] W 45cm. From 16.4 by smaller size, narrower wingbar, shallower tail notch and less distinct white bases to flight feathers; difficult to separate from 16.5 but wingbar narrower and tail notch less deep. ✳ Inshore, offshore and open ocean. Nests in burrows. ☉ Ha.

16.2 FORK-TAILED STORM-PETREL *Oceanodroma furcata* [Océanite à queue fourchue] W 46cm. No other similar-sized bird at sea. ✳ Offshore and open ocean. ☉ Ha.

16.3 LEACH'S STORM-PETREL *Oceanodroma leucorhoa* [Océanite culblanc] W 46cm. Rump varies between being streaked (**a**) and all-white (**b**). White-rumped morph very similar to 16.6 but wings longer and flight more erratic. ✳ Offshore and open ocean. Nests in burrows on slopes with boulders, grass or trees. ♫ *tsjek*, followed by a prolonged, slowly rising chatter. ☉ Breeds N Pacific, W N. America.

16.4 MATSUDEIRA'S STORM-PETREL *Oceanodroma matsudairae* [Océanite de Matsudaira] W 56cm. Note prominent white shafts at base of primaries. ✳ Open seas. ☉ Breeds Japan.

16.5 TRISTRAM'S STORM-PETREL *Oceanodroma tristrami* [Océanite de Tristram] W 56cm. From 16.4 by lack of white patch at base of primaries and perhaps less distinct wingbar. Note large, long-winged jizz. ✳ Offshore and open ocean. Nests in burrows or crevices or under clumps of vegetation. ♫ High, soft, rhythmic barking. ☉ Breeds Ha 8,10,11,13,16.

16.6 BAND-RUMPED (or Madeiran) **STORM-PETREL** *Oceanodroma castro* [Océanite de Castro] W 45cm. Very similar to 16.3, but white of rump reaches further down to lower underparts and wings are shorter. ✳ Offshore and open ocean. Nests in burrows and crevices. ☉ Breeds Ha 2,7. R

16.7 SOUTH GEORGIA DIVING-PETREL *Pelecanoides georgicus* [Puffinure de Géorgie du Sud] W 32cm. At sea not safely separable from 16.8; in hand by less mottling to throat and neck, whiter underwing coverts, paler inner webs of primaries and more obvious white tips to scapulars. Both species fly low, fast and with whirring wings ✳ Offshore and open ocean. Nests in narrow burrows in open or sparsely vegetated terrain. ♫ Series of small, rising, hoarse shrieks. ☉ Breeds NZ 12.

16.8 COMMON DIVING-PETREL[21] *Pelecanoides urinatrix* [Puffinure plongeur] W 36cm. See 16.7. ✳ Inshore, also offshore. Nests in burrows under vegetation, but also in the protection of boulders. ♫ Soft, rising moans. ☉ Main breeding sites found in NZ 1,4–6,8.

17 TROPICBIRDS, GANNETS & BOOBIES

17.1 RED-BILLED TROPICBIRD *Phaethon aethereus* [Phaéton à bec rouge] W 103cm. Ad. from 17.2–3 by barred back; Imm. from Imm. 17.3 (also with yellow, dark-tipped bill) by denser back barring and more distinct black outer wings. ❋ Open seas. Breeds on cliff faces. ⊙ FrPo 2,3, Ma, Ha 18. R

17.2 RED-TAILED TROPICBIRD *Phaethon rubricauda* [Phaéton à brins rouges] W 112cm. White-backed Ad. unmistakable. Imm. shows dark bill and lengthwise striped flight feathers (not forming a more or less solid patch). ❋ Open seas. Breeds on cliff faces. ♪ Low or high, hoarse, rather shrill croaks and tern-like shrieks. ⊙ NZ 10, Ha, Co, Fi, FrPo, Pi.

17.3 WHITE-TAILED TROPICBIRD *Phaethon lepturus* [Phaéton à bec jaune] W 93cm. Upperwings distinctly patterned; cf. 17.1 for Imm. ❋ Open seas, but can be seen inshore. Nests on cliff faces and in trees. ♪ Very high, rasping, often staccato shrieks. ⊙ Breeds most island groups in the area. R

17.4 AUSTRALIAN (or ^{NZ}Australasian) **GANNET** *Morus serrator* [Fou austral] W 165cm. Reaches Ad. plumage in 3 years. Juv., Imm. and Ad. plumages shown. Ad. from Ad. 17.5 by white outer tail feathers and shorter gular line; from 17.7 by facial pattern and white inner secondaries. ❋ Normally inshore. Breeds on cliffs and small rocky islands. ♪ High, gull-like shrieking. ⊙ Breeds mainland and nearby islands NZ 1,2,4,6,7.

17.5 CAPE GANNET *Morus capensis* [Fou du Cap] W 170cm. See 17.4. ❋ Offshore. ⊙ Once NZ 1.

17.6 ABBOTT'S BOOBY *Sula abbotti* [Fou d'Abbott] W unknown. Underwings mainly white. Distinctive large-headed jizz. Juv. like Ad., but black parts very dark brown. ❋ Offshore and open seas. Breeds Christmas I. ⊙ NMa 4.

17.7 MASKED BOOBY *Sula dactylatra* [Fou masqué] W 165cm. Ad. from 17.4–5 by absence of yellow wash to head and by black inner secondaries. Underwings of Imm. paler than underwings of Imms 17.8–9. ❋ Offshore. Breeds exposed on low-lying small islands. ♪ Low, short, hoarse barks and gull-like shrieks. ⊙ Breeds many island groups in the area. R

17.8 BROWN BOOBY *Sula leucogaster* [Fou brun] W 140cm. Upperparts solid dark brown. Shown are ssp *plotus* (**a**, W & C Pacific) and ssp *etesiaca* (**b** Imm., CE Pacific, which shows pale head and typical brown mottling of underparts. ❋ Offshore and open seas. Variety of habitats, from small low-lying islands to cliffs on larger islands. ♪ Low, Mallard-like cackling. ⊙ Breeds many island groups in the area.

17.9 RED-FOOTED BOOBY *Sula sula* [Fou à pieds rouges] W 143cm. Polymorph, from almost all-white (**a**) to all-brown (not shown); **b** is the pale brown-headed morph. All Ad. morphs show red feet and pink-based, blue bill. ❋ Open ocean. Nests on beaches of small islands and in trees and scrub of larger islands. ♪ Very low, hoarse croaking. ⊙ Breeds most island groups in the area.

18 FRIGATEBIRDS, PELICAN & CORMORANTS

18.1 GREAT FRIGATEBIRD (or Iwa) *Fregata minor* [Frégate du Pacifique] W 215cm. From smaller 18.2 by black axillaries, though Imms in NW Hawaiian islands may show some irregular white feather tips in axillaries. ✳ Offshore and open seas. Nests in mangrove and bushes. ♪ Very high, hurried *djibdibdjib*— or high, hoarse, drawn-out shrieks. ☉ Widespread in the area, but rare at NZ. R

18.2 LESSER FRIGATEBIRD *Fregata ariel* [Frégate ariel] W 185cm. Note white spur across axillaries. ✳ Offshore and open seas. Nests in mangrove and bushes. ♪ High, hoarse twittering and quacking *bicbicbic*—. ☉ Widespread in the area, but rare near NZ and Ha. R

18.3 AUSTRALIAN PELICAN *Pelecanus conspicillatus* [Pélican à lunettes] W 240cm. Unmistakable, very large bird. ✳ Large open water bodies, inland and at coast. ♪ Low, fast, slightly pig-like rattling. ☉ NZ 2, Fi, Pa, Na.

18.4 DARTER[22] (or [AOU]Anhinga) *Anhinga melanogaster* [Anhinga roux] W 120cm. Unmistakable by white stripe from below eyes, long kinked neck and neatly arranged, lengthwise striped scapulars and wing feathers. Often swims with submerged body. ✳ Sheltered, shallow waters (partly) with fringing trees or dense bush. ♪ Angry, duck-like cackling. ☉ NZ 1,2, ?Pa.

18.5 GREAT CORMORANT[23] *Phalacrocorax carbo* [Grand Cormoran] W 130cm. Note white throat and cheeks; Br plumage shows white thighs, a small orange triangle below eyes and often short white streaks on crown and upper neck. Imm. from other cormorants and shags by large size and yellow at gape. ✳ Large coastal and inland waters, but occasionally also on smaller water bodies like pools. ♪ Low *wohwoh - -*. ☉ NMa 6, Mi 4, NZ 1–5,7,10.

18.6 LITTLE BLACK CORMORANT (or [NZ]Shag) *Phalacrocorax sulcirostris* [Cormoran noir] W 100cm. Note slender bill, all-dark plumage (without any white, except some spots and streaks at head in Br plumage) and scalloped upperparts. Gregarious. ✳ Wetlands and sheltered coastal waters. ☉ NZ 1–3,?4,10, Fi 3, Pa.

18.7 LITTLE PIED CORMORANT *Microcarbo melanoleucos*[24] [Cormoran pie] W 88cm. Polymorph with varying degree of white to underparts from (almost) all-black (**b**) to white-throated (**a**) or all white below (**c**). Long-tailed! Note short, blunt, yellow bill. ✳ Any inland or coastal water from pools to estuaries. ♪ Soft barks in irregular series. ☉ Mi 4, NMa 1,6, Pa, NZ 1–7.

18.8 PIED CORMORANT (or [NZ]Shag) *Phalacrocorax varius* [Cormoran varié] W 120cm. Large. Yellow bare skin in front of blue eyes diagnostic. Gape pink, yellow in Imms. ✳ Subcoastal and inshore waters. Prefers sites with mangroves, some trees or high bush, but also with large rocks and cliffs. ♪ High, hoarse shrieks *ark ark*. ☉ NZ 1–3,5.

18.9 PELAGIC CORMORANT *Stictocarbo pelagicus* [Cormoran pélagique] W 101cm. Note very small size, very slender jizz with long, thin bill and red skin around eyes. White flanks missing in N-br plumage. ✳ Inshore and offshore. Breeds N Pacific. ☉ Ha 13,17.

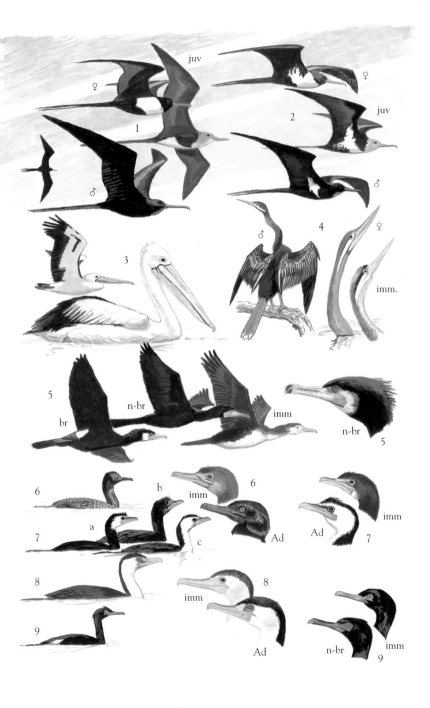

19 SHAGS[25]

19.1 SPOTTED SHAG[26] *Phalacrocorax punctatus* [Cormoran moucheté]
W 95cm, L 69cm. From 19.9 by range and white stripe from eye down neck
(faintly so in N-br plumage). ❋ Estuaries, inshore and offshore waters.
♫ Very high, rhythmic shrieks and low, pig-like grunts. ☉ NZ 1–3,5. E.NZ

19.2 BRONZE (or [NZ]Stewart Island) **SHAG**[27] *Phalacrocorax chalconotus*
[Cormoran bronzé] L 68cm. Polymorph, pied (**a**), bronze (**b**) and intermediate
morph (**c**) shown. Best diagnosed on basis of range. Note also the facial pattern,
especially of bare parts. Shows a white dorsal patch. ❋ Islands off mainland
coast. ♫ High, shivering cries. ☉ NZ 2,3. E.NZ

19.3 ROUGH-FACED (or [NZ]New Zealand King) **SHAG** *Phalacrocorax
carunculatus* [Cormoran caronculé] L 76cm. Best diagnosed on basis of range.
Note also the facial pattern, especially of bare parts. ❋ Islands off mainland
coast. ♫ Low grunts and high cries. ☉ NZ 2. R E.NZ

19.4 CHATHAM ISLANDS SHAG *Phalacrocorax onslowi* [Cormoran des
Chatham] L 63cm. Best diagnosed on basis of range. Note also the facial pattern,
especially of bare parts. Shows a white dorsal patch. ❋ Inshore. ♫ Low, hollow
grunts. ☉ NZ 4. E.NZ

19.5 CAMPBELL ISLANDS SHAG *Phalacrocorax campbelli* [Cormoran de
Campbell] L 63cm. Best diagnosed on basis of range. Note also the facial pattern,
especially of bare parts, and that foreneck is black. ❋ Inshore and offshore.
☉ NZ 7,8. R E.NZ

19.6 BOUNTY ISLANDS SHAG *Phalacrocorax ranfurlyi* [Cormoran de
Bounty] L 71cm. Best diagnosed on basis of range. Note also the facial pattern,
especially of bare parts. ❋ Inshore ☉ NZ 5,9. E.NZ

19.7 AUCKLAND ISLANDS SHAG *Phalacrocorax colensoi* [Cormoran des
Auckland] L 63cm. Best diagnosed on basis of range. Note also the facial pattern,
especially of bare parts, and that black may be closed at foreneck as shown in
perched bird or open as shown in flying bird. Shows a white dorsal patch.
❋ Inshore and offshore. ♫ Very low, hollow grunts. ☉ NZ 5,6. E.NZ

[19.8 MACQUARIE ([NZ]Island) **SHAG** *Leucocarbo purpurascens* [Cormoran de
Macquarie] L 75cm. Best diagnosed on basis of range. Note also the facial
pattern, especially of bare parts. ❋ Inshore. ♫ Very low croaks. ☉ Occurs only at
and near Macquarie I., which is Australian Territory and therefore outside the
area.]

19.9 PITT ISLAND SHAG *Stictocarbo featherstoni* [Cormoran de Featherston]
L 63cm. From 19.1 by solid black head and neck. ❋ Inshore and offshore.
♫ High, hoarse shrieks. ☉ NZ 4. E.NZ

20 EGRETS & HERONS

20.1 PACIFIC HERON *Ardea pacifica* [Héron à tête blanche] L 91cm. Unmistakable by size and contrast between grey body and white neck and head. In flight, when seen head-on, striking white 'head-lights' at wrists. ✴ Shallow water of wetlands and inundated grassland, swamps, watercourses, irrigation channels, ponds. Occasionally in estuaries. ♫ High, hoarse, drawn-out shrieks. ☉ NZ 1,2,10.

20.2 GREAT EGRET *Ardea alba* [Grande Aigrette] L 92cm. From 20.3 by range. Legs of American N-br ssp *egretta* black as shown; otherwise plumage like 20.3. In courtship plumage (not shown, but probably never seen in the area) with red bill. ✴ Marsh, lake edges, river margins, estuaries, mudflats. ♫ Series of high, cackling croaks. ☉ Ha. R

20.3 EASTERN GREAT EGRET[28] *Ardea modesta* L 92cm. Upper legs in N-br plumage yellowish as shown. Note blackish bill tip in Br plumage. In courtship plumage (**cs**) with black bill and red legs (reversed in 20.2). ☉ Pa, NMa 1, Mi 4, Fi 1,3, NZ 1,2,4–7.

20.4 WHITE-FACED HERON *Egretta novaehollandiae* [Aigrette à face blanche] L 67cm. Note white head and brown-purplish breast. ✴ Inland and coastal wetlands, estuaries, rivers, lakes and farmland. ♫ Chicken-like, hoarse cackling. ☉ NZ 1,2,4–7,10, A.Sa 4, Ton 12, Fi 1,28, Ni.

20.5 LITTLE EGRET *Egretta garzetta* [Aigrette garzette] L 60cm. 3 ssps might be involved: Nom. (**a**, with all-yellow toes, NW Pacific), *immaculata* (**b**, with black, yellow-soled toes, NZ) and *nigripes* (**c**, with all-black toes). Bill of N-br Nom. is grey, including lores; lores of all other ssps and stages is yellow and bill is black with varying pale at base of lower mandible. ✴ Inland and coastal wetlands, estuaries, mudflats, fresh- and saltmarsh, floodplains. ♫ Low, unmusical, indignant shrieks. ☉ NZ 1,2,10, Pa, Mi 2,4, NMa 1–4, Gu.

20.6 PACIFIC REEF-HERON *Egretta sacra* [Aigrette sacrée] L 62cm. Polymorph, white (**a**), dark (**b**) and intermediate (**c**) morphs shown. Strong, short, green legs diagnostic. Most dark-phase birds have small white spot at throat as shown (**d**). ✴ Rocky shorelines, coral reefs, estuaries, mangroves, sandy and muddy beaches. ♫ Mid-high, hoarse, angry shrieks. ☉ Widespread, but not in Ha and at NZ 8,9.

20.7 LITTLE BLUE HERON *Egretta caerulea* [Aigrette bleue] L 63cm. From 20.6 by habitat, range, much more slender jizz and longer legs. Note small black tips to primaries in flight (difficult to see). Colour morphs are age-related (not so in 20.6). ✴ Any fresh or saline water with some fringing vegetation. ☉ Ha 6. V

20.8 SNOWY EGRET *Egretta thula* [Aigrette neigeuse] L 58cm. Note bright yellow lores and toes of Ad. (shown in flight). Imm. has dark legs with yellowish stripe at back. In Br plumage from Br 20.5 by many plumes at head. ✴ Beaches, mudflats, swamps. ☉ ?A.Sa, Ha. R

20.9 INTERMEDIATE (or [NZ]Plumed) **EGRET** *Egretta intermedia* [Héron intermédiaire] L 63cm. Nom. (**a**, N and W Pacific) and race *plumifera* (**b**, NZ) shown. N-br plumage for both ssps similar, except yellowish upper legs of **a**, which are all-black for **b** as shown in flight; bill of Br **a** is yellow with black culmen (not shown) but all yellow in **b**; courtship bill (**cs**) is black in **a**, while reddish, yellow-tipped in **b**. ✴ Wetlands, floodplains, swamps, watercourses. ♫ Mumbled croaks and low, short, hoarse chatters. ☉ Pa, NMa 1–4,6, Mi 3,4, Gu, NZ 1,2, Ha. R

20.10 CATTLE EGRET[29] *Bubulcus ibis* [Héron garde-boeufs] L 51cm. Nom. (**a**, I Ha) and race *coromandus* (**b**, elsewhere in Pacific), differing in extent of apricot colour in Br plumage. Unmistakable by jizz, small size and habits. Rarely/never seen standing in water. Imms might show blackish bill and legs. ✴ Usually found in association with cattle and other domestic stock. ♫ Hoarse croaking in colony. ☉ NZ 1,2,4,5,10, Fi 1, Pa, NMa, Mi 2–4, Ma, Ha, Fr Po 2.

21 HERONS & BITTERNS

21.1 BLACK-CROWNED NIGHT-HERON (or Aukuu) *Nycticorax nycticorax* [Bihoreau gris] L 59cm. Ad. unmistakable by colour pattern and stocky jizz. Imm. from Imm. 21.2 by orange (not yellow) eyes and less heavy streaking. ❋ Wooded and forested edges of fresh and saline water bodies. ♫ Dry, low, sustained irregular *oh-oh - oh-*. ☉ NMa 1,2, Gu, Pa, Mi 2–4, Ha 1–7.

21.2 RUFOUS NIGHT-HERON[30] *Nycticorax caledonicus* [Bihoreau cannelle] L 60cm. Ad. unmistakable by colour pattern and stocky jizz. Imm. (and Ad.) show yellow eyes. Saline and freshwater bodies with fringing forest, mangroves, estuaries; also in flooded grassland, swamps and reedbeds. ♫ Scratchy, hoarse shrieks. ☉ NZ 1,2, NMa 6, Mi 3,4, Pa. R

21.3 JAPANESE NIGHT-HERON *Gorsachius goisagi* [Bihoreau goisagi] L 49cm. See 21.4. ❋ Rivers and swamps in forest. ☉ Pa 2. V

21.4 MALAYAN NIGHT-HERON *Gorsachius melanolophus* [Bihoreau malais] L 49cm. From 21.3 by black streak through crown, less regular pattern on underparts and differences in wing pattern. ❋ Streams and marsh in rainforest. ☉ Pa.

21.5 SCHRENCK'S BITTERN *Ixobrychus eurhythmus* [Blongios de Schrenck] L 36cm. From 21.6 by dark chestnut upperparts and different wing pattern. ❋ Marsh, reedbeds, swamps, rice paddies. ☉ Pa. V

21.6 YELLOW BITTERN *Ixobrychus sinensis* [Blongios de Chine] L 35cm. Note uniform-coloured upperparts with black flight feathers. ❋ Marshes, swamps, flooded fields, dense vegetation at water bodies; sometimes far from water. ☉ Gu, Mi 3,4, Pa, NMa 1–4.

21.7 LITTLE BITTERN[31] *Ixobrychus minutus* [Blongios nain] L 32cm. Unmistakable by white wing patch. ❋ Normally in freshwater marshes with reedbeds. ♫ Series of very low grunting *roh* notes. ☉ NZ 2. V

21.8 BLACK BITTERN *Ixobrychus flavicollis* [Blongios à cou jaune] L 60cm. Unmistakable by dark upperparts with yellowish streak through cheeks and purplish neck and breast streaking. ❋ May be found in forest at edge of streams, dense woodland, ponds, estuaries, rank grassland. ☉ Gu. R

21.9 AUSTRALASIAN BITTERN *Botaurus poiciloptilus* [Butor d'Australie] L 71cm. Secretive. Note cryptic colouring with distinctively tawny cheeks. ❋ Wetlands with tall, dense vegetation. Occasionally in lagoons or estuaries. ♫ Series of extreme low, well-separated, resounding *boom* notes. ☉ NZ 1–3.

22 HERONS & IBISES

22.1 GREY HERON[32] *Ardea cinerea* [Héron cendré] L 94cm. From 22.2 by range, size and lack of chestnut in plumage. ✳ Normally at the edge of any salt, saline or freshwater body. ♫ Call: sudden *wráh*. ☉ Fi 1, Mi 4, Pa, NMa 1,4.

22.2 GREAT BLUE HERON *Ardea herodias* [Grand Héron] L 115cm. Note size and chestnut in plumage. All-white morph (**a**, from 20.2 by bulkier jizz and pale legs) has not been seen in the area. ✳ At the edge of any fresh or saline water. ♫ Call: low, raucous *fraaah*. ☉ Ha.

22.3 CHINESE POND-HERON *Ardeola bacchus* [Crabier chinois] L 45cm. Contrast, suddenly exposed when flying up, between white wings and rest of body plumage diagnostic. ✳ Normally in swamps, at riverbanks, rice paddies, mangroves; occasionally also in dry grassland. ☉ Gu. V

22.4 GREEN HERON *Butorides virescens* [Héron vert] L 40cm. Irregular guest from America. Ad. from 22.5 by range and by dark chestnut neck and breast sides. Imm. not safely separable but by range. ✳ Dense vegetation at shallow waters. ☉ Ha.

22.5 STRIATED HERON *Butorides striata* [Héron strié] L 40cm. See 22.4. Shown is ssp *amurensis* (Pa, NMa, Mi 3,4, with grey back and neck sides). Not shown are ssp *diminuta* (Fi, Ton, possibly NZ 10, with grey-brown back and pale grey underparts, streaked pale rufous) and ssp *patruelis* (Tahiti, with brown back and tawny-orange cheeks and neck sides). ✳ Mainly in coastal habitats such as mangroves, estuaries, bare tidal mudflats. Occasionally away from coast. ♫ Generally silent. ☉ Pa, NMa 1, Mi 3,4, Fi, Ton 12, NZ 10, FrPo 4.

22.6 STRAW-NECKED IBIS *Threskiornis spinicollis* [Ibis d'Australie] L 65cm. Distinctive black wings and white body and tail. ✳ Wetlands, dry or wet grassland. ☉ NZ 2. V

22.7 AUSTRALIAN IBIS[33] *Threskiornis molucca* [Ibis à cou noir] L 70cm. Unmistakable by white plumage, red line across underwings and short tail. ✳ Swamps, wet grassland, recently burnt vegetation, mudflats, shores, estuaries, mangroves, saltmarsh. ♫ Unmusical, hoarse shrieks. ☉ NZ 1,2.

22.8 WHITE-FACED IBIS *Plegadis chihi* [Ibis à face blanche] L 56cm. Slender, dark bird with distinctive facial pattern. From 22.9 by range. ✳ Marshes, rice paddies, margins of freshwater bodies; occasionally at saline waters or in dry areas. ☉ Ha.

22.9 GLOSSY IBIS *Plegadis falcinellus* [Ibis falcinelle] L 56cm. From 22.8 by range and facial pattern. ✳ Shallow waters, flood plains, shallow lake margins, pools, occasionally in estuaries or on tidal mudflats. ♫ Mid-high, hoarse shrieks *euheuheuh—*. ☉ Fi, NZ 1,2,4.

23 SPOONBILLS, WHISTLING-DUCK, SWANS, SHELDUCKS & GOOSE

23.1 ROYAL SPOONBILL *Platalea regia* [Spatule royale] L 78cm. Unmistakable. Note black bill and legs. ✳ Large shallow water bodies. Estuaries, tidal mudflats, occasionally in lagoons. May also be seen in inland wetlands. ♪ Low hoarse grunts. ☉ NZ 1,2,4,10.

23.2 YELLOW-BILLED SPOONBILL *Platalea flavipes* [Spatule à bec jaune] L 88cm. Unmistakable. Note yellow bill and legs. ✳ Normally in inland wetlands, rarely coastal. ♪ High, loud, strident *wheech* notes. ☉ NZ 1. V

23.3 PLUMED WHISTLING-DUCK[34] *Dendrocygna eytoni* [Dendrocygne d'Eyton] L 50cm. Unmistakable by upright stance, pale plumage and elongated flank plumes. ✳ Normally in grassland, often near wetland. ♪ Very high whistles like *feeeuw*. ☉ NZ 1,2. V

23.4 AUSTRALIAN (or [NZ]Chestnut-breasted) **SHELDUCK** *Tadorna tadornoides* [Tadorne d'Australie] L 66cm. Unmistakable by dark plumage with chestnut breast. ✳ Grassland, wetlands, lagoons, estuaries. ♪ Mid-high, nasal or pumped, rather toneless quacking. ☉ NZ 1,2,4–6,10.

23.5 PARADISE SHELDUCK *Tadorna variegata* [Tadorne de paradis] L 65cm. No similar bird in range. White head of ♀ especially distinctive. ✳ Farmland, pastures, dams, pools, mountain streams. ♪ Mewing, drawn-out quacks. ☉ NZ 1,2,4,10. E.NZ

23.6 CAPE BARREN GOOSE *Cereopsis novaehollandiae* [Céréopse cendré] L 83cm. Unmistakable by shape and colour of bill and grey plumage. ✳ Normally in grassland and wetlands. ♪ Very low, pig-like or higher-pitched, duck-like quacking. ☉ NZ 1,2. I

23.7 TUNDRA SWAN *Cygnus columbianus* [Cygne siffleur] L 128cm. Nom. shown (**a**, with mainly or all black bill) and ssp *bewickii* (**b**, with yellow-and-black bill; note that yellow of bill is at base, not at tip of bill, cf. 23.8). ✳ Winters on grassland and marshes. ☉ Ha 2,5,7,17.

23.8 MUTE SWAN *Cygnus olor* [Cygne tuberculé] L 142cm. Unmistakable. Neck held more gracefully bent than 23.7. ✳ Small lakes, ponds, wetlands, streams. ♪ Hissing. ☉ NZ 1,2. I

23.9 BLACK SWAN *Cygnus atratus* [Cygne noir] L 125cm. No similar bird known. Note white outer wing in flight. ✳ Fresh and saline water bodies with submergent or fringing vegetation. Grazes also in adjacent pastures. ♪ Irregular series of high oboe-like notes. ☉ NZ 1,2,4.

24.1 SNOW GOOSE *Chen caerulescens* [Oie des neiges] L 75cm. Unmistakable by black primaries. 'Blue' morph (**a**, shown to smaller scale) not yet recorded in the tropical Pacific. ✳ Winters normally on farmland near W coast of America. ♪ Very high cackling. ☉ Ha, Ma 17. V

24.2 EMPEROR GOOSE *Chen canagica* [Oie empereur] L 78cm. Black foreneck and barred body plumage distinctive. ✳ Winters normally along rocky sea coasts. ☉ Ha.

24.3 GREATER WHITE-FRONTED GOOSE *Anser albifrons* [Oie rieuse] L 76cm. Dark bars across belly and white bill base (both missing in Imms) diagnostic. ✳ Winters normally in open, treeless areas. ♪ Very high cackling. ☉ Ha.

24.4 GREYLAG GOOSE *Anser anser* [Oie cendrée] L 80cm. Note large size and (partly) pale upperwings. ✳ Open farmland, lakes, estuaries. ☉ NZ 1,2,4. I

24.5 MANED (Goose or [NZ]Australian Wood) **DUCK** *Chenonetta jubata* [Canard à crinière] L 50cm. Note 3 parallel stripes over back and thin-necked jizz. ✳ Short-grassed areas near smaller water bodies often surrounded by woodland or forest. ♪ Low, grumbled shrieks in series: *wreeék - -*. ☉ NZ 1,2,5.

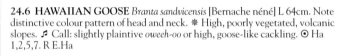

24.6 HAWAIIAN GOOSE *Branta sandvicensis* [Bernache néné] L 64cm. Note distinctive colour pattern of head and neck. ✳ High, poorly vegetated, volcanic slopes. ♪ Call: slightly plaintive *oweeh-oo* or high, goose-like cackling. ☉ Ha 1,2,5,7. R E.Ha

24.7 BRANT *Branta bernicla nigricans* [Bernache cravant] L 61cm. Very dark, no colours in plumage, only black, grey and white. ✳ Winters normally in estuaries and grassland near coast. ♪ Rather low cackling. ☉ Ha.

24.8 CANADA GOOSE *Branta canadensis* [Bernache du Canada] L 65–100cm (size depending on race). All ssps larger than 24.9. Four ssps: *occcidentalis* (**a**, Dusky, darkest ssp, Irr); *parvipes* (**b**, Lesser, note thin upper neck); *moffitti* (**c**, Moffitt's, note long-necked jizz); *maxima* (**d**, Giant, with white marks above eyes). ✳ Great variety of habitats, mainly near water bodies. ♪ Well-known, high goose-honks. ☉ Ha, NZ 1,2,5,6,10. Also once seen Ki 14.

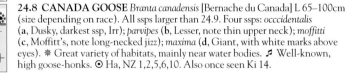

24.9 CACKLING GOOSE *Branta hutchinsii* [Bernache de Hutchins] L 55–65cm. (Much) smaller than 24.8. Four ssps: *minima* (**a**, Cackling, may show small white spot at base of neck); Nom. (**b**, Richardson's, white neck ring, Irr); *leucopareia* (**c**, Aleutian, paler than **b**, but not safely separable, Irr); *taverneri* (**d**, Taverner's, note short bill and long thick neck). ✳ Winters normally near sea coast. ☉ Ha.

25.1 BLUE DUCK *Hymenolaimus malacorhynchos* [Canard bleu] L 53cm. Unmistakable in its range and habitat. Well camouflaged when perching on and between boulders, but note striking yellow eyes and pale bill. ✳ Fast mountain streams in forested areas. ♫ Low, grating *wrrrreh* (of ♀) and very high *feeeh* (of ♂). ☉ NZ 1,2. R E.NZ

25.2 PINK-EARED DUCK *Malacorhynchus membranaceus* [Canard à oreilles roses] L 40cm. Unmistakable by barred flanks and relatively large bill. ✳ Normally on shallow water. ♫ Very high, liquid, fluted chattering. ☉ NZ 1. V

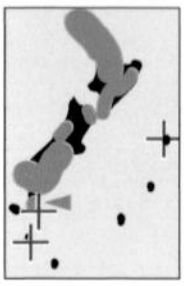

25.3 AUSTRALIAN (or [NZ]Australasian) **SHOVELER**[35] *Anas rhynchotis* [Canard bridé] L 50cm. Br ♂ easily separable from Br ♂ 25.4 by facial pattern and scaled, not all-white breast and upper mantle; ♀ from ♀ 25.4 by greenish dusky bill, missing orange tones at gape. N-br ♂ difficult to separate from N-br ♂ 25.4; both may show ghost of pale moon sickle between eye and bill, but N-br ♂ 25.4 tends to be whiter on lower breast. ✳ Large wetlands, floodplains and lakes. Occasionally on saline waters. ♫ Irregular *quack* notes. ☉ NZ 1–6.

25.4 NORTHERN SHOVELER *Anas clypeata* [Canard souchet] L 48cm. See 25.3. ✳ Freshwater bodies, estuaries and lagoons. ♫ Low *duckduck*. ☉ Ha, Ki 1,2,3, NMa, Mi 1,2,4, Wake I, Ma 1, NZ 1,2

25.5 CHESTNUT TEAL *Anas castanea* [Sarcelle rousse] L 45cm. N-br ♂ ♂ and esp. ♀ ♀ very similar to 25.6, but chin more buffy, less whitish. From 25.7 by red, not dark-brown eyes, lack of eyering and different upper- and underwing pattern. ✳ Saline and coastal water bodies. ☉ NZ 1,2.

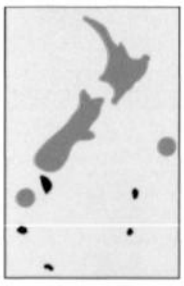

25.6 GREY TEAL *Anas gracilis* [Sarcelle australasienne] L 43cm. Note pattern of upper- and underwings, similar to that of 25.5. Throat whitish. ✳ Shallow inland and coastal water bodies. ♫ E.g. low grunted quacks or high chatters. ☉ NZ 1,2,4,5.

25.7 BROWN TEAL *Anas chlorotis* [Sarcelle de Nouvelle-Zélande] L 48cm. From 25.5 by dark eyes, white eyering and different wing pattern. ✳ Estuaries and wetlands with open water. ☉ NZ 1,2. E.NZ

25.8 AUCKLAND ISLANDS TEAL *Anas aucklandica* [Sarcelle d'Auckland] L 48cm. From 25.7 by very restricted range. Flightless. ✳ At pools and creeks in moorland. ♫ Duet-like, fast chatter, partly sizzling and partly quacking. ☉ NZ 6. E.NZ

25.9 CAMPBELL ISLAND TEAL *Anas nesiotis* [Sarcelle de Campbell] L 40cm. From 25.7 by very restricted range and small size. Flightless. ✳ Gullies and pools on tussock slopes. ♫ Long, high, shivering twitters. ☉ NZ 7,12. R E.NZ

26 DUCK, WIGEONS, TEALS, GADWALL, PINTAIL & GARGANEY

26.1 EURASIAN WIGEON *Anas penelope* [Canard siffleur] L 48cm. Br ♂ unmistakable. Note white belly, greyish underwing, black speculum with pale inner secondaries and grey-black bill pattern of ♀. N-br ♂ like ♀, but flanks rustier. ❋ Prefers in winter coastal habitats such as estuaries, lagoons and brackish marshes. ♪ Low *grr-grr*; very high up-and-down *weeéeew*. ☉ NMa, Pa, Mi 3,4, Ma 11, Ha, Wake I., Palmyra I.

26.2 AMERICAN WIGEON *Anas americana* [Canard d'Amérique] L 51cm. Br ♂ unmistakable; ♀ and N-br ♂ like 26.1 but flanks (even) rustier and underwing whiter. ❋ Winters in coastal wetlands. ♪ Growl; very high mellow whistles *whec whec*. ☉ Ha, NMa 1, Gu, Palmyra I.

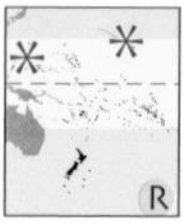

26.3 GADWALL *Anas strepera* [Canard chipeau] L 51cm. Br ♂ grey with black rear end. N-br and ♀ show whiter belly than Mallard. Note pattern of upperwing with partly white speculum (very restricted in Imm. ♀). ❋ Wetlands with shallow water. In winter also in coastal habitats. ♪ Very low nasal *creck* (♂), low *quack* (♀). ☉ NMa 1,2, Ha.

26.4 EURASIAN TEAL *Anas crecca* [Sarcelle d'hiver] L 36cm. ♀ like ♀ 26.5, but cheeks often plainer without dark smear. Br ♂ distinctive by head and rear-end pattern. N-br ♂ and ♀ with rather plain face sides and little orange at gape. ❋ Shallow water bodies; in winter also in coastal habitats. ♪ Like 26.5. ☉ NMa, Pa, Mi 4, Ki 3, Ha, Ma.

26.5 GREEN-WINGED TEAL[36] *Anas carolinensis* [Sarcelle à ailes vertes] L 36cm. Br ♂ like 26.4 but green at head more narrowly bordered white, no white scapulars and with vertical white stripe at breast sides. ♀ similar to ♀ 26.4, but often with dark smear to cheeks (though some ♀♀ 26.4 show this feature too). ❋ Like 26.4. ♪ Toy trumpet-like *eeeh-eh-eh*; very high short fluted *peek peeh*. ☉ Ma 11, Ha, Ma.

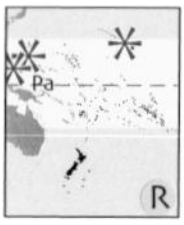

26.6 GARGANEY *Anas querquedula* [Sarcelle d'été] L 39cm. Note distinctive white eyebrow of Br ♂; N-br ♂ and ♀ show broad white trailing edge to wing, strong head pattern with whitish spot near bill base. ❋ Winters on open freshwater lakes and coastal lagoons. ♪ Dry rattle *crrrruh* (♂) and high, loud, hoarse cackling (♀). ☉ Ha, Pa, NMa 1,2, Gu.

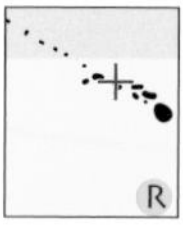

26.7 BAIKAL TEAL *Anas formosa* [Sarcelle élégante] L 41cm. Br ♂ unmistakable; N-br ♂ and ♀ show broken eyebrow and white spot at bill base. ❋ In winter on fresh and brackish water bodies, including flood plains and estuaries. ☉ Ha 7.

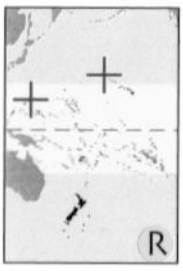

26.8 FALCATED TEAL *Anas falcata* [Canard à faucilles] L 51cm. N-br ♂ and ♀ show plain dark grey-brown head and all-black bill. ❋ In winter on fresh and saline water bodies. ☉ NMa 1, Ha 17. V

26.9 NORTHERN PINTAIL *Anas acuta* [Canard pilet] L 60cm (♂), 55cm (♀). Note slim jizz with elegant curved neck. N-br ♂ from ♀ by bill pattern. ❋ Winters in open wetlands, estuaries and lagoons. ♪ Very high, croaking *wrrik-wrrik-wrik*. ☉ Ha, Wake I., Ma, NMa, Pa, Mi, Ki 3, Ton, Co, FrPo 1,3.

27 MALLARD, DUCKS & TEALS

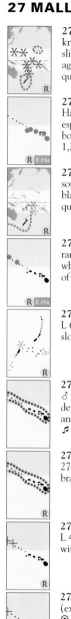

27.1 MALLARD *Anas platyrhynchos* [Canard colvert] L 58cm. Br ♂ very well known and distinctive; note orange, partly dusky bill of ♀. N-br ♂ like ♀ but slightly darker and with greenish bill. ❋ Wetlands, estuaries, settled and agricultural areas. ♫ Well-known, loud quacking from ♀ and low, soft, grunted quacks from ♂. ☉ Ha, NZ, NMa, Mi, Ki, Ma, Co, Tu, Fi. I?

27.2 HAWAIIAN DUCK (or Koloa-maoli) *Anas wyvilliana* [Canard des Hawaï] L 51cm. Resembles ♀ Mallard, but smaller, darker and with rufous tinge, especially below. Speculum more green, less blue than 27.1. ❋ Any freshwater body. ♫ Mallard-like vocalisations but slightly higher and thinner. ☉ Ha 1,2,6,7,8. E.Ha

27.3 PACIFIC BLACK (or [NZ]Grey) **DUCK** *Anas superciliosa* [Canard à sourcils] L 57cm. Note very dark plumage with striking facial pattern and all-black bill. ❋ Any habitat with some shallow water. ♫ Series of Mallard-like quacks. ☉ Pa 1, Mi 3, Fi, Sa, Ton, WaF, Co 6, FrPo 1,29, NZ (Not in Ha).

27.4 LAYSAN DUCK *Anas laysanensis* [Canard de Laysan] L 41cm. Restricted range. Very small and dark. No clear distinction between ♂ and ♀, amount of white to face sides variable in both sexes (see **a** and **b**). Prefers to walk. ❋ Lagoon of Laysan, from where wanders thr. ☉ Ha 13,17. E.Ha

27.5 SPOT-BILLED DUCK *Anas poecilorhyncha* [Canard à bec tacheté] L 61cm. Yellow bill tip diagnostic. ❋ Normally on freshwater lakes, marshes and slow-moving rivers. ☉ Gu.

27.6 BLUE-WINGED TEAL *Anas discors* [Sarcelle à ailes bleues] L 38cm. Br ♂ distinctive. N-br ♂ (not shown) and ♀ like ♀ 27.7 but with more clearly defined facial pattern and plumage markings, less warm colouring of underparts and less distinctively spatulated bill. ❋ Winters on brackish, coastal waters. ♫ Very high *tjip* (♂), yelping *wec* (♀). ☉ Ha.

27.7 CINNAMON TEAL *Anas cyanoptera* [Sarcelle cannelle] L 42cm. See 27.6. Bill larger and more spatulated than 27.6. ❋ Prefers shallow fresh and brackish water bodies. ♫ Low cackling and grumbling. ☉ Ha.

27.8 HARLEQUIN DUCK *Histrionicus histrionicus* [Arlequin plongeur] L 45cm. Note head pattern and winter habitat. ❋ Winters along rocky coasts with rough water. ☉ Ha 13,17.

27.9 LONG-TAILED DUCK *Clangula hyemalis* [Harelde kakawi] L 43cm (excl tail). Very distinctive. Note winter habitat. ❋ Winters in offshore waters. ☉ Ha 1,6,17.

28 DUCKS, SCAUPS, POCHARD & REDHEAD

28.1 (^NZ^Australian) **WHITE-EYED DUCK** (or Hardhead) *Aythya australis* [Fuligule austral] L 53cm. From 28.2 by different bill pattern, white vent and larger white area on underparts. ✳ Deep fresh and brackish waters. Also in marshes and mountain lakes. ♪ Very low, rasping quacks. ⊙ NZ 1,5.

28.2 NEW ZEALAND SCAUP *Aythya novaeseelandiae* [Fuligule de Nouvelle-Zélande] L 40cm. Normally unmistakable by dark plumage (head and neck blackest) and all-grey, black-tipped bill. Note white at bill-base of ♀. Restricted pale area on underparts in flight. ✳ Open freshwater lakes. ♪ High, shivering whistles and low quacks. ⊙ NZ 1,2. E.NZ

28.3 RING-NECKED DUCK *Aythya collaris* [Fuligule à collier] L 42cm. Peaked head diagnostic. Note bill pattern of ♂ and narrow white eyering of ♀; crescent in front of ♀ eye may be white or rufous-brown as shown. Flight feathers rather grey, not white. From similar black-backed 28.7 by red eyes and lack of elongated crest feathers. The reddish-brown ring around the neck (for which this species is named) is usually inconspicuous and hardly visible in the field. ✳ Fresh and saline water bodies. ⊙ Ha.

28.4 CANVASBACK *Aythya valisineria* [Fuligule à dos blanc] L 53cm. Very distinctive jizz with peaked crown. Note all-black bill. ✳ Larger, deep water bodies. ⊙ Ha, Ma.

28.5 COMMON POCHARD *Aythya ferina* [Fuligule milouin] L 50cm. ♂ from rather similar 28.6 by less grey to bill and paler body plumage; ♀ crown less peaked than 28.4, but more so than 28.6. Note also bill shape. ✳ Larger water bodies, lagoons, estuaries. ⊙ Gu, NMa 1, Ha 17.

28.6 REDHEAD *Aythya americana* [Fuligule à tête rouge] L 48cm. Yellow eyes of ♂ diagnostic. ♀ is rather uniform grey brown (not with more contrasting head as shown by 28.4–5). ✳ Winters on large, quiet, fresh and saline water bodies. ♪ High *théréthéréthérewhere* or Mallard-like *wehwehweh*. ⊙ Ha.

28.7 TUFTED DUCK *Aythya fuligula* [Fuligule morillon] L 43cm. Long crest (shorter in ♀) diagnostic. Most ♀♀ like **a**, however, some with varying amount of white at bill base (**b**), but this not sharply demarcated as in crestless ♀ 28.8. ✳ Open water of any size, slow rivers, estuaries, coastal bays. ⊙ Pa, Mi 4, NMa, Ma 1, Ha.

28.8 GREATER SCAUP *Aythya marila* [Fuligule milouinan] L 45cm. Main differences with slightly smaller 28.9 are whiter flanks and more white in wings of ♂♂; ♀ and ♂ with longer bill. Normally with rounded head, not slightly peaked at rear crown like 28.9. ✳ Winters mainly in brackish and salt coastal waters. ⊙ Ha, Mi 4, NMa 1.

28.9 LESSER SCAUP *Aythya affinis* [Petit Fuligule] L 43cm. See 28.8. ✳ Winters on brackish waters; occasionally on sea. ⊙ Ha.

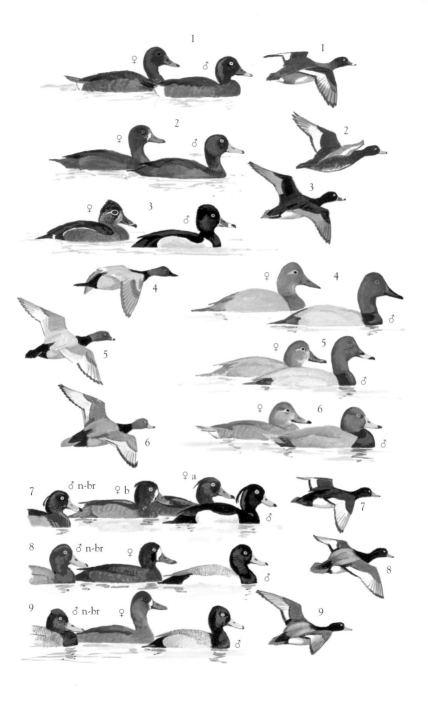

29 GOLDENEYES, MERGANSERS, SCOTERS & BUFFLEHEAD

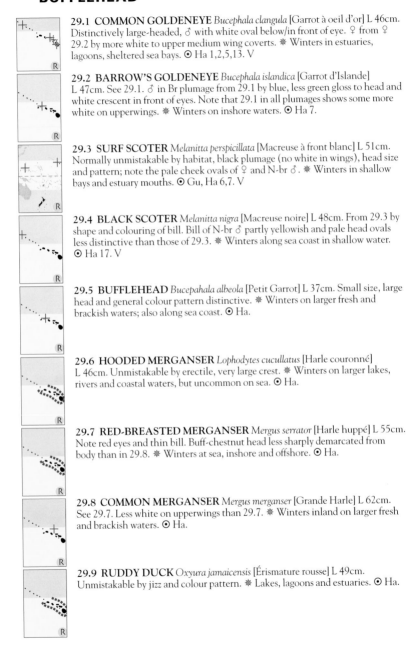

29.1 COMMON GOLDENEYE *Bucephala clangula* [Garrot à oeil d'or] L 46cm. Distinctively large-headed, ♂ with white oval below/in front of eye. ♀ from ♀ 29.2 by more white to upper medium wing coverts. ✳ Winters in estuaries, lagoons, sheltered sea bays. ☉ Ha 1,2,5,13. V

29.2 BARROW'S GOLDENEYE *Bucephala islandica* [Garrot d'Islande] L 47cm. See 29.1. ♂ in Br plumage from 29.1 by blue, less green gloss to head and white crescent in front of eyes. Note that 29.1 in all plumages shows some more white on upperwings. ✳ Winters on inshore waters. ☉ Ha 7.

29.3 SURF SCOTER *Melanitta perspicillata* [Macreuse à front blanc] L 51cm. Normally unmistakable by habitat, black plumage (no white in wings), head size and pattern; note the pale cheek ovals of ♀ and N-br ♂. ✳ Winters in shallow bays and estuary mouths. ☉ Gu, Ha 6,7. V

29.4 BLACK SCOTER *Melanitta nigra* [Macreuse noire] L 48cm. From 29.3 by shape and colouring of bill. Bill of N-br ♂ partly yellowish and pale head ovals less distinctive than those of 29.3. ✳ Winters along sea coast in shallow water. ☉ Ha 17. V

29.5 BUFFLEHEAD *Bucepahala albeola* [Petit Garrot] L 37cm. Small size, large head and general colour pattern distinctive. ✳ Winters on larger fresh and brackish waters; also along sea coast. ☉ Ha.

29.6 HOODED MERGANSER *Lophodytes cucullatus* [Harle couronné] L 46cm. Unmistakable by erectile, very large crest. ✳ Winters on larger lakes, rivers and coastal waters, but uncommon on sea. ☉ Ha.

29.7 RED-BREASTED MERGANSER *Mergus serrator* [Harle huppé] L 55cm. Note red eyes and thin bill. Buff-chestnut head less sharply demarcated from body than in 29.8. ✳ Winters at sea, inshore and offshore. ☉ Ha.

29.8 COMMON MERGANSER *Mergus merganser* [Grande Harle] L 62cm. See 29.7. Less white on upperwings than 29.7. ✳ Winters inland on larger fresh and brackish waters. ☉ Ha.

29.9 RUDDY DUCK *Oxyura jamaicensis* [Érismature rousse] L 49cm. Unmistakable by jizz and colour pattern. ✳ Lakes, lagoons and estuaries. ☉ Ha.

30 EAGLES & OSPREY

30.1 OSPREY *Pandion haliaetus* [Balbuzard pêcheur] W 158cm. Ssp *carolensis* (**a**, without breast collar, see perched birds) and Nom. (**b**, with collar, see flying bird) shown. Unmistakable by general jizz and colour pattern. Plunge-dives after fish (**c**). ✳ At any larger water body that should be relatively calm and clear. ☉ Ha, NMa 1,6, Pa, Mi 4, Gu. R

30.2 GOLDEN EAGLE *Aquila chrysaetos* [Aigle royal] W 210cm. Large raptor with rather long tail and bulging-out trailing edge of wings. Note 'golden' nape and vent. Ad. with pale lesser upperwing coverts; Imm. with striking white in tail and wings. ✳ Normally in open mountain areas. ☉ Ha 7. V

30.3 WHITE-TAILED EAGLE[37] *Haliaeetus albicilla* [Pygargue à queue blanche] W 215cm. White tail of Ad. in combination with irregular patterned pale brown body plumage diagnostic. Imm. from Imm. 30.4 by relatively shorter tail and less massive bill. ✳ At large water bodies, from sea coast to large marshes. ☉ Ha 7. V

30.4 STELLER'S SEA EAGLE *Haliaeetus pelagicus* [Pygargue empereur] W 206cm. Ad. unmistakable; cf. Imm. with Imm. 30.3. ✳ Coasts, large lagoons and large lakes. ☉ Ha 11,17,18. V

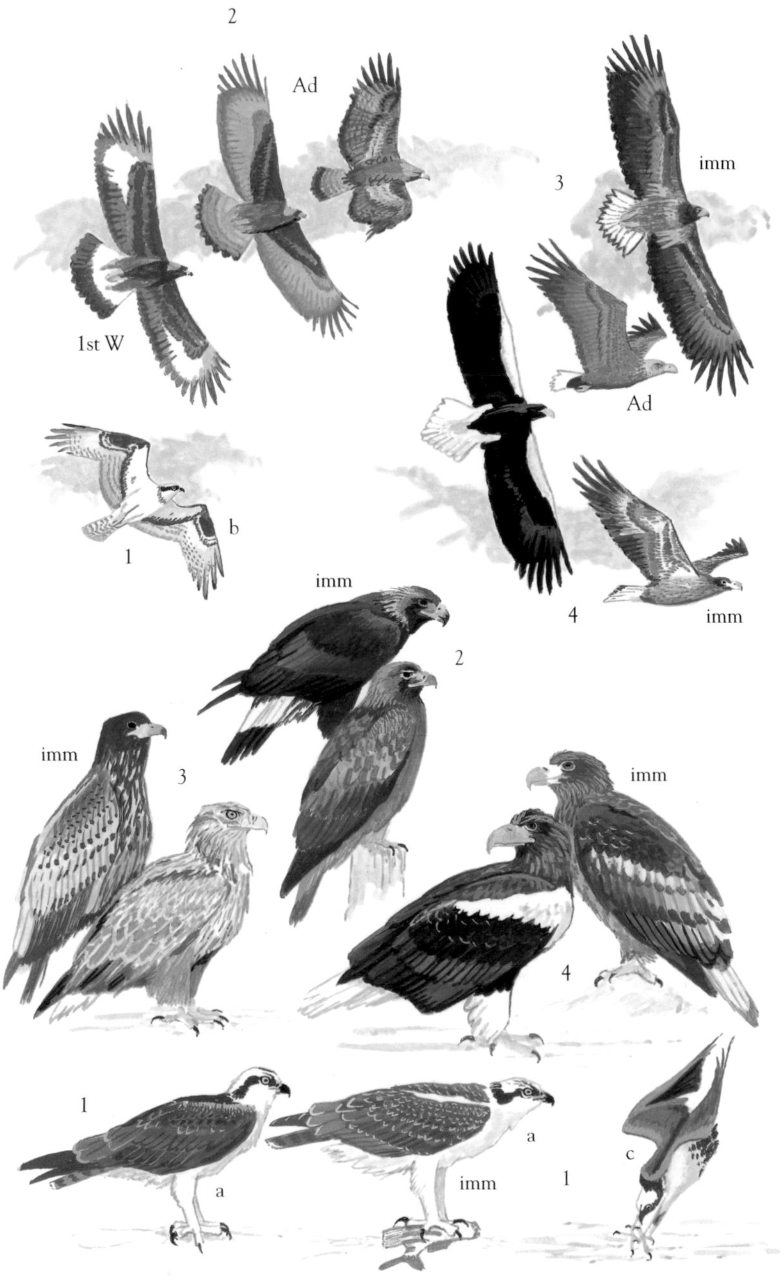
2
Ad
1st W
3
imm
Ad
b
1
imm
4
imm
2
imm
3
imm
4
1
a
a
imm
1
c

31 HARRIERS, SPARROWHAWK & GOSHAWK

31.1 NORTHERN HARRIER *Circus cyaneus* [Busard Saint-Martin]
W 110cm. So far only seen on Hawaii I.; ♂ from ♂ 3 by black streaks on head, mantle and upperwing coverts. ♀ and Imm. from 31.3 by more pronounced white mark on lower rump and different underwing coverts (underwing of Imm. and ♀ 31.1 is basically overall creamy white with some dark bars across flight feathers, while underwing of 31.3 is basically dark with a pale patch, formed by base of primaries). ❋ Flat or gently sloping country, grassland, moorland, wetlands, <2,500m. ♫ Very high, sharp yells and mewing. ☉ Ha.

31.2 SWAMP HARRIER *Circus approximans* [Busard de Gould] W 131cm. Large raptor, with wings characteristically held in shallow V. No other harrier species in range, while other raptors in NZ are only rare 33.5 and vagrants 33.3 and 33.4. ❋ Wetlands, grassland, <1,700m. ♫ High, sharp, mewing *vweeét* notes. ☉ NZ 1–7,10, Co 1, Ton, Fi.

31.3 WESTERN MARSH-HARRIER *Circus aeruginosus* [Busard des roseaux] W 120cm. Vagrant to NMa. ♂ unmistakable; for ♀ and Imm. cf. 31.1. ❋ Marsh with extensive reedbeds, open wetlands, estuaries. ☉ NMa 4. V

31.4 JAPANESE SPARROWHAWK *Accipiter gularis* [Épervier du Japon] L 27cm (♂), 30cm (♀). Note overall barring of underwings of ♂, ♀ and Imm. ♂ eye is red, not black like ♂ 31.5, while underparts are barred (not distinctly so), while those of 31.5 are uniform pale orange-buff. ❋ Forest and woodland. ☉ ?Pa.

31.5 CHINESE GOSHAWK (or ^AOU^Gray Frog-Hawk) *Accipiter soloensis* [Épervier de Horsfield] L 28cm (♂), 35cm (♀). Cf. 31.4; note orange cere of ♂ and black wingtips of ♂ and ♀. ❋ Forest and woodland near more open areas. ☉ Ha 18, Mi 4, NMa 1,4, Pa.

31.6 FIJI GOSHAWK *Accipiter rufitorques* [Autour des Fidji] L 35cm (♂), 40cm (♀). ♀ like ♂. Sole goshawk in Fiji and no other raptors except 31.2 and 33.7 (both with very different jizz). Note pale underparts of Ad. without markings. ❋ Forest, woodland, city parks. ☉ Fi thr., excl. 7. E.Fi

32.1 BRAHMINY KITE *Haliastur indus* [Milan sacré] W 117cm. Ad. unmistakable; cf. underwing pattern of Imm. with 32.2, which can look very similar but latter with forked tail. ❋ Normally at coast and islands, but can be seen inland in a great variety of habitats, including refuse dumps. ☉ Pa. V

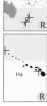

32.2 BLACK KITE *Milvus migrans* [Milan noir] W 145cm. Distinctive forked tail and long, kinked wings. ❋ Very varied, including settlement and cities, but not in dense forest. ♫ Very high, quavering, slow, trilled *pihurrrrrrrr* (*pi* very short). ☉ Pa, NMa 1,6, Ha 17, NZ 1,2.

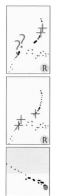

32.3 ROUGH-LEGGED BUZZARD (or [AOU]Hawk) *Buteo lagopus* [Buse pattue] L 55cm. Very variable, darker or paler morphs possible, most with dark belly; white rump and tail base diagnostic, but this feature not present in all individuals. Most ♀♀ with single dark subterminal tail band, most ♂♂ with double band. Hovers frequently. ❋ Normally in open country, marsh. ☉ Ha 3,13,17. V

32.4 EURASIAN BUZZARD *Buteo buteo* [Buse variable] L 55cm. Very variable, most with distinctive dark wrist patches (but these normally less dark than in 32.3). Shown in flight ssp *toyoshima*. ❋ Forest edges and half-open and open areas with scattered trees or posts for perching. ♫ In flight very high, mewing, rapidly descending, whistled *níau*. ☉ NMa 5,9, ?Pa.

32.5 GREY-FACED BUZZARD *Butastur indicus* [Busautour à joues grises] L 46cm. Note throat stripe, barred belly and long straight wings. ❋ Forest near more open areas. ☉ Mi 4, Gu, Pa.

32.6 HAWAIIAN HAWK *Buteo solitarius* [Buse d'Hawaï] L 45cm. Pale (**a**) and all-dark (**b**) morphs occur in equal numbers. Compact jizz with short, rounded wings. ❋ From farmland to forest. Up to 2,700m. ♫ High, drawn-out *eeeoh* or series of excited *kiew* notes. ☉ Ha 1. R E.Ha.

33 KESTRELS, FALCONS, MERLIN & PEREGRINE

33.1 EURASIAN KESTREL *Falco tinnunculus* [Faucon crécerelle] L 36cm. Very similar to 33.3 but not seen in same range. Hunting method (diving from hovering flight) distinctive. Note typical falcon 'tear' from eye corner and long tail with black/dark subterminal bar. ❋ Any habitat with scattered trees or posts for perching, except forest interior. ♫ High, sharp *keekeekeekee*—. ⊙ Gu, NMa 1. V

33.2 AMUR FALCON *Falco amurensis* [Faucon de l'Amour] L 29cm. Note diagnostic red bare parts and orange lower belly and vent. Underwing coverts of ♂ are white, those of ♀ and Imm. barred. ❋ Open areas with some trees or tree clumps. ⊙ NMa 1. V

33.3 AUSTRALIAN (or ^NZ^Nankeen) **KESTREL** *Falco cenchroides* [Crécerelle d'Australie] L 33cm. Like 33.1, but smaller, paler, less spotted and striped and not in same range. ❋ Open or wooded areas, open forest, towns. ♫ Slightly irregular series of *ketsch* notes. ⊙ NZ 1,2. R

33.4 BLACK FALCON *Falco subniger* [Faucon noir] L 51cm. Unmistakable by uniform dark plumage and large size. ❋ Grassland and open woodland. ⊙ NZ 1,2. V

33.5 NEW ZEALAND FALCON *Falco novaeseelandiae* [Faucon de Nouvelle-Zélande] L 45cm. Unmistakable in its range (easily separable from Swamp Harrier, which is larger with different jizz and flight with longer wings, held in shallow V). ❋ Mosaic of forest and grassland, pastures and tussock land, <2,100m. ♫ Sharp, very high, hurried series *weetweetweet*-. ⊙ NZ 1–3,6,7. R E.NZ

33.6 MERLIN *Falco columbarius* [Faucon émerillon] L 38cm. From 33.7 by smaller size, quicker wingbeats and less distinctive moustachial stripes. ♂ orangey below. ❋ Normally winters in open areas and estuaries. ⊙ Ha. V

33.7 PEREGRINE FALCON *Falco peregrinus* [Faucon pèlerin] L 40cm (♂), 50cm (♀). Three ssps in area possible, *calidus* (not shown, very similar to **a**, V NMa), *japonicus* (**a**, wanderer to W and N Pacific) and *nesiotis* (**b**, much darker with broader moustachial stripe, E. to Fiji). ❋ Hunts over any habitat but needs cliffs (or, as replacement, the front of buildings) for nesting. ⊙ NMa 1, Gu, Mi 4, Pa, Ha, Fi.

34 CRANES, SCRUBFOWLS, TURKEY, GUINEAFOWL & PARTRIDGES

34.1 BROLGA *Grus rubicunda* [Grue brolga] L 130cm. From 34.2 by range, head pattern, darker primaries and narrow dark trailing edge to wings. ❀ Normally found in wetlands and wet grassland. ♬ Excited, guttural trumpeting. ☉ NZ 1,2. V

34.2 SANDHILL CRANE *Grus canadensis* [Grue de Canada] L 115cm. Cf. 34.1 ❀ Normally in open fields and meadows. ♬ High, upslurred rattle *urrr-rrrr*. ☉ Ha 6. V

34.3 NIAUFOOU SCRUBFOWL *Megapodius pritchardii* [Mégapode de Pritchard] L 33cm. Very dark with pale head. ❀ Sloping, forested areas, especially on innerside of volcano calderas. ☉ Ton 7. E.Ton R

34.4 MICRONESIAN SCRUBFOWL *Megapodius laperouse* [Mégapode de La Pérouse] L 29cm. Paler than 34.3 with grey underparts; ranges far apart. ❀ Forest, forest remains, coconut groves, coastal scrub. ☉ Pa, NMa. R

34.5 WILD TURKEY *Meleagris gallopavo* [Dindon sauvage] L 122cm (♂), 93cm (♀). Unmistakable. Feral birds (more slender than farm birds) are probably interbred ssps from N America and Mexico. ❀ Prefers mixture of farmland, grassland and wooded plots. ♬ Well-known, high gobbling. ☉ Ha 1,2,4,5,8, Fi 4,7,8, NZ 1,2. I

34.6 HELMETED GUINEAFOWL *Numida meleagris* [Pintade de Numidie] L 68cm. Unmistakable. ❀ Feral in agricultural areas. ☉ NZ 1. I

34.7 CHUKAR (or ^NZ^Chukor) *Alectoris chukar* [Perdrix choukar] L 32cm. From 34.8 by sharply defined collar that frames upper breast, more regular black flank barring and brown restricted to scapulars. Grey crown separated from bill by rather broad black band that connects eyes. ❀ Steep rocky hillsides and mountain slopes. ♬ Dry, rapid, rising cackling, often in chorus. ☉ NZ 1,2, Ha 1,2,4,5,7. I

34.8 RED-LEGGED PARTRIDGE *Alectoris rufa* [Perdrix rouge] L 36cm. From 34.7 by different flank barring, brownish neck, black streaking running from collar and grey of crown touching bill. ❀ Prefers dry, hilly country with scattered bush. ☉ NZ 1,2. I

34.9 GREY PARTRIDGE *Perdix perdix* [Perdrix grise] L 30cm. Note tawny-rufous face sides, rufous flank barring and streaked and barred upperparts. ❀ Large, grassy areas with some scattered, dense scrub. ♬ Series of anxious *tok* notes, interrupted by *kreeh-èh*. ☉ ?NZ. I

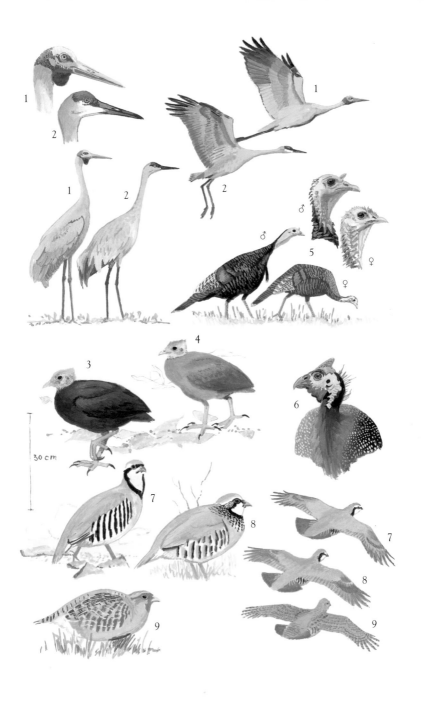

35.1 ([AOU]Northern) **BOBWHITE** ([NZ]Quail) *Colinus virginianus* [Colin de Virginie] L 25cm. Unmistakable by facial pattern and barred underparts. Note grey wings and tail in flight. ❋ Normally in open, brushy country, forest edge, roadsides. ♫ High, well-separated, fluting *bob-Whéet*. ☉ ?NZ. I

35.2 CALIFORNIA QUAIL *Callipepla californica* [Colin de Californie] L 25cm. Confusion only possible with 35.3. Latter has rufous crown, dark reddish flanks and dark belly patch. Note also forehead, which is paler in this species. ❋ Mosaic of open, tussocky places, scrub and taller cover. ♫ Very high, rapid twittering or toneless, staccato ticking. ☉ NZ 1,2,4, Ha 1,2,4,5,7. I

35.3 GAMBEL'S QUAIL *Callipepla gambelii* [Colin de Gambel] L 25cm. See 35.2. ❋ Like 35.2. ♫ Peacock-like *pwéew*; very high descending *swéar*. ☉ Ha 1,3,4. I

35.4 BROWN QUAIL *Coturnix ypsilophora* [Caille tasmane] L 20cm. Very plain; barred dark, especially below and striped overall cream. Sexes alike. ❋ Shrub- and grassland with tall, rank vegetation. ♫ Very high, irregular, fluting *pree-peep* or *tjut vwweeét*. ☉ Fi 1,2,19, NZ 1. I

35.5 JAPANESE QUAIL *Coturnix japonica* [Caille du Japon] L 19cm. From 35.4 by range and lack of barring below. ❋ Open areas with short, grassy vegetation. ♫ Series of excited dry *tarátaah*. Call is a typical quail-like *pruweet*. ☉ Ha. I

35.6 BLUE-BREASTED QUAIL *Coturnix chinensis* [Caille peinte] L 13cm. Very small; ♂ unmistakable; barring below of ♀ interrupted at breast and belly. ❋ Grass- and shrubland with short dense vegetation. ☉ Gu. I

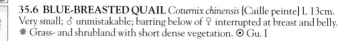

35.7 ERCKEL'S FRANCOLIN *Francolinus erckelii* [Francolin d'Erckel] L 41cm. Large size, rufous crown and overall striping diagnostic. ❋ In the transition zone between grassland and open forest. Often at higher altitudes than 35.8 and 35.9. ♫ Dry cackling. ☉ Ha. I

35.8 GREY FRANCOLIN *Francolinus pondicerianus* [Francolin grise] L 34cm. Buffy head, striking barring and rufous outer tail feathers diagnostic. ❋ Dry grassland and scrub, often at road sides, golf courses and lawns. ☉ Ha. I

35.9 BLACK FRANCOLIN *Francolinus francolinus* [Francolin noir] L 35cm. ♂ Unmistakable; ♀ shows chestnut neck patch and delta-shaped body feathers. ❋ Areas with tall grass and cultivation. ♫ Very high, very dry shrieks. ☉ Ha 1–7, Gu. I

36.1 RING-NECKED (or [NZ]Common) **PHEASANT** *Phasianus colchicus* [Faisan de Colchide] L 82cm (♂), 57cm (♀). ♂ unmistakable by general colour pattern and neck ring. ♀ from ♀ 36.2 by less intense marking especially on upperparts. ✳ Grassland, farmland, heath, scrub, forest edge, cultivation. ♫ High, dry *u'uhuuh*, followed by wing flutter. ☉ Ha 1,2,6,7, NZ 1,2.

36.2 GREEN PHEASANT *Phasianus versicolor* [Faisan versicolore] L 82cm (♂), 58cm (♀). ♂ from ♂ 36.1 by darker plumage and lack of neck ring. ✳ Open woodland, bush, farmland. ☉ Ha. I (but might be extirpated).

36.3 KALIJ PHEASANT *Lophura leucomelanos* [Faisan leucomèle] L 68cm (♂), 55cm (♀). Mix of Nom. and ssp *hamiltonii*. ♂ unmistakable. Note crest, scalloped plumage and 2-toned tail of ♀. ✳ Forest with dense undergrowth, abandoned cultivation. ♫ Call: high to very high, rapid *prutprutprut—*. ☉ Ha 1. I

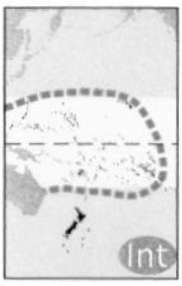

36.4 RED JUNGLEFOWL *Gallus gallus* [Coq bankiva] L 70cm (♂), 44cm (♀). Original wild Junglefowl (unmistakable) introduced on many islands in Pacific by Polynesians; where it still occurs it interbreeds with farm chickens. ✳ Forest edge and interior, secondary growth. ♫ Unmistakable *e-nuh-e-nuh*. ☉ Widespread, but not in NZ. I

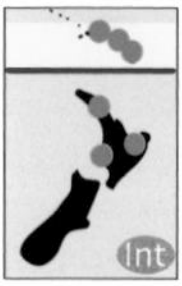

36.5 INDIAN PEAFOWL *Pavo cristatus* [Paon bleu] L 205cm (♂), 74cm (♀). Unmistakable. ✳ Forest, woodland, cultivation with hedgerows and glades. ♫ Well-known *MIAUUH*. ☉ Ha, NZ 1,?2. I.

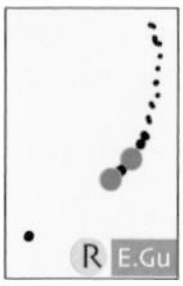

36.6 GUAM RAIL *Gallirallus owstoni* [Râle de Guam] L 28cm. Distinctive, dark rail with unbanded upperparts except barred flight feathers. Flightless. ✳ Varied habitats from short grass, fern thickets, woodland to forest edge. ☉ Gu (reintroduced), I NMa 4. E.Gu

36.7 BUFF-BANDED RAIL[39] *Gallirallus philippensis* [Râle tiklin] L 30cm. From 36.6 by different coloured head pattern, pink bill and legs and markings on upperparts. Several ssps, all showing buff breast band. ✳ Any type of wetland, including swamp, marsh, sides of creeks, rivers and lagoons, farm dams, damp heathland. ♫ High, slightly upslurred, shrieking *vreeet* or very high, staccato *whuut* and high, sharp *sweep-sweep-sweep*. ☉ NZ 1–3,6, Ni, Sa, Ton, Fi, Pa, WaF.

36.8 WEKA[40] *Gallirallus australis* [Râle wéka] L 55cm (♂), 48cm (♀). Flightless. Four ssps recognised, differing in brightness and contrast of colours and intensity of streaking; of these, Nom. and *scotti* have morphs (grey, chestnut and black). As example, head of black morph Nom. shown (**a**). ✳ Margins of any type of wetland, water or forest with some low cover, including sea beach and cultivated land. ♫ High, upslurred shrieks or high, liquid *wew* or ascending series of *cohweétcohwéet*. ☉ NZ 1–4. E.NZ.

36.9 AUCKLAND ISLANDS RAIL[41] *Lewinia muelleri* [Râle d'Auckland] L 23cm. Note compact jizz and distinctive, darkish colour pattern. ✳ Damp, densely vegetated areas with a (sub-)canopy at 1m above ground. ♫ High *frueeh*. ☉ Restricted to 2 islands off NZ 6. E.NZ

37.1 BAILLON'S CRAKE[43] *Porzana pusilla* [Marouette de Baillon] L 18cm. Note green bill, red eyes and barred undertail coverts. ✸ Any type of wetland, normally with dense or clumped vegetation. ♫ Low, short, dry rattle, decelerating just at the end. ☉ NZ 1–4.

37.2 SORA RAIL *Porzana carolina* [Marouette de Caroline] L 22cm. Distinctive facial pattern. Note yellow bill and partly yellow undertail coverts. Not seen in range of 37.1. ✸ Normally in wetlands, especially marshes with shallow water and floating and submergent vegetation. ♫ Swept-up *oh what* or very high rather plaintive very fast *ohwhatagainthinbickering*. ☉ Ha 1,6. V

37.3 WHITE-BROWED CRAKE *Porzana cinerea* [Marouette grise] L 18cm. Distinctive facial pattern. ✸ Any type of wetland, especially those with floating vegetation and at rivers and creeks, but also in cultivated areas. ☉ Mi 2,3,4, Fi 1,10,11, Pa 1,2,5,6, Ma 3, Sa 1,2.

37.4 HENDERSON ISLAND CRAKE *Porzana atra* [Marouette de Henderson] L 18cm. Uniform velvety black. ✸ Forest, thickets, coconut groves. ♫ Drawn-out churring note, lowered at the end. ☉ Pi 1. R E.Pi

37.5 SPOTLESS CRAKE *Porzana tabuensis* [Marouette fuligineuse] L 17cm. Two-toned: grey below, brown above. ✸ Densely vegetated wetlands, cultivated areas, rice paddies, fern-covered hillsides, forest. ♫ Series of well-separated *wih* notes or a mid-high rattle. ☉ Mi 1, Fi 1,3,4,10,11,12, Sa 1, A.Sa 2, Ton, ?Ni, Co 2,3,9, FrPo, Pi 4, NZ 1–4,10.

37.6 RED-LEGGED CRAKE *Rallina fasciata* [Râle barré] L 24cm. From 37.7 by colour of legs and spotted outer wing. ✸ Reedy wetlands, wet areas in forest. ☉ Pa. V

37.7 SLATY-LEGGED CRAKE *Rallina eurizonoides* [Râle de forêt] L 22cm. Note uniform, unmarked wings. ✸ Dense grassy areas in forest; occasionally in mangrove. ☉ Pa 1–6.

37.8 PLAIN BUSH-HEN *Amaurornis olivacea* [Râle des Philippines] L 28cm. Unmistakable by dark plumage, green bill and long-legged and -necked jizz. ✸ Swamps and adjoining cultivation and forest edges. ☉ Pa 8. V

1

imm

2

br

♀

n-br

imm

3

5

4

6

imm

8

7

imm

38 WATERCOCK, MOORHENS, SWAMPHEN, TAKAHE & COOTS

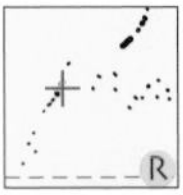

38.1 WATERCOCK *Gallicrex cinerea* [Râle à crête] L 42cm (♂), 36cm (♀). Unmistakable by colour pattern, size and upright stance. Shown is Br ♂ , N-br ♂ like ♀. ✳ Normally in swamps and wet cultivation. ☉ Pa.

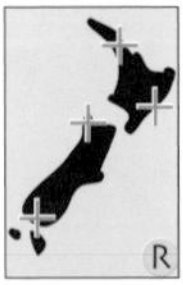

38.2 BLACK-TAILED NATIVE-HEN *Gallinula ventralis* [Gallinule aborigène] L 34cm. Unmistakable by colour pattern and tail shape. ✳ Straggles from Australia to NZ and can then be found anywhere, even in streets and gardens. ♪ Very short, dry and squeaky rattles. ☉ NZ 1,2. R

38.3 DUSKY MOORHEN *Gallinula tenebrosa* [Gallinule sombre] L 37cm. From somewhat smaller 38.4 by more uniform-coloured plumage, lack of white line along side of body and differently coloured legs. ✳ Normally in wetlands with open water, bordered by some fringing vegetation. May forage on adjoining short grass. ♪ Series of well-separated, staccato *wic* and *úweeh* notes. ☉ NZ 2. R

38.4 COMMON MOORHEN *Gallinula chloropus* [Gallinule poule-d'eau] L 34cm. Note orange 'garters'. ✳ Wide range of mainly freshwater wetlands, normally with access to some open water. ♪ High, nasal, irregular *tic tic - -*. ☉ Ha 6,7,17, NMa 1,2,4, Mi 4, Pa 1,6, Gu.

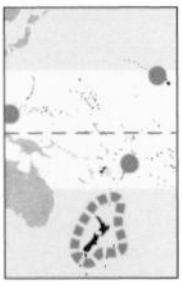

38.5 PURPLE SWAMPHEN[44] *Porphyrio porphyrio* [Talève sultane] L 46cm. Unmistakable by size, long legs and wings and by jizz. Shown is black and blue ssp *melanotus* ([NZ]Pukeko). Ssps *pelewensis* (Palau) and *samoensis* (Samoa and Fiji) may show much green and brown above and cerulean-blue on upper breast. ✳ Wetlands and adjoining open habitats such as grassland, cultivation, lawns, sports fields. ♪ Peacock-like *uweeéh*. ☉ NZ 1–4,7,10, Fi, Ni, Ton, Sa, Pa 1,2,5,6.

38.6 SOUTHERN (or [NZ]South Island) **TAKAHE** *Porphyrio mantelli* [Talève takahé] L 63cm. Flightless. Unmistakable. ✳ Alpine tussocky grassland and scrubland. ♪ Very high, yelping *Weeh!-Weeh!-Weeh!* in duet. ☉ Introduced on some small, predator-free islands off NZ 1,2. R E.NZ

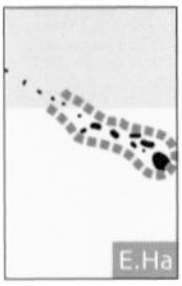

38.7 HAWAIIAN COOT *Fulica alai* [Foulque des Hawaï] L 39cm. Note absence of black pointed wedge between frontal shield and upper mandible. Undertail coverts white. 2 colour morphs: **a** with all white bill and shield and **b** yellow, red or cream shield. Both morphs may have subterminal ring on bill as shown for **b**. ✳ Wide variety of wetlands, including estuaries, marshes, ponds, flooded land. ♪ High, staccato *bic bic bic - -*. ☉ Ha 1–7. R E.Ha

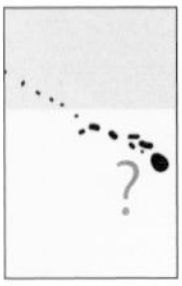

38.8 AMERICAN COOT *Fulica americana* [Foulque d'Amérique] L 38cm. Note absence of black, pointed wedge between frontal shield and upper mandible. Undertail coverts white. Shows dark red subterminal bill ring and swelling at top of shield. ✳ Normally on open lakes, canals and other calm water with overhanging vegetation. ♪ Sharp, bouncing *bic* or *wruk wwurriet*. ☉ Ha; several records, but identification unproven.

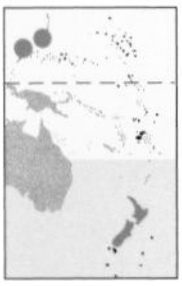

38.9 EURASIAN COOT[45] *Fulica atra* [Foulque macroule] L 37cm. Black wedge between frontal shield and upper mandible and no white at undertail coverts. ✳ Wetlands with open water, well furnished with submergent vegetation. ♪ Coughing *weh* and high staccato *bic* or yelping *uweh uweh uweh*. ☉ NZ 1,2, NMa 2, Gu.

39 OYSTERCATCHERS, STILTS & AVOCET

39.1 VARIABLE OYSTERCATCHER *Haematopus unicolor* [Huîtrier variable] L 48cm. Polymorph, all-black (**a**) and intermediate (**b**) morphs unmistakable; pied morph (**c**) is very similar to 39.4, but shows in flight a square-cut, black or smudgy lower back, while wingbars are narrow, not touching trailing edge. ✳ Prefers sandy coasts, especially near estuaries. ♪ High *tuweet*, like 39.4 but more plaintive and gull-like. ☉ NZ 1–3. E.NZ *Note*: shown is a group of ♂ ♂ displaying while uttering a shrill, rattling call.

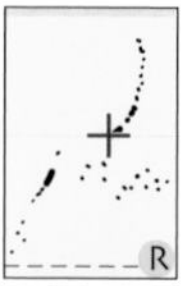

39.2 EURASIAN OYSTERCATCHER *Haematopus ostralegus* [Huîtrier pie] L 44cm. Note long bill and wide wingbars extending over bases of primaries. Sole oystercatcher in the area with a partial white collar in N-br plumage. ✳ Sandy and rocky beaches, mudflats, estuaries, agricultural land. ♪ Very high, very sharp, resounding *bicbicbic—*. ☉ Gu.

39.3 CHATHAM (NZIsland) **OYSTERCATCHER** *Haematopus chathamensis* [Huîtrier des Chatham] L 48cm. Normally the sole oystercatcher in the Chatham archipelago, from where it does not wander to mainland or other islands. 39.5 has been recorded from Chatham I., but that species shows a deep white wedge on back. Note relatively short bill. ✳ Prefers rocky shores. ♪ Like 39.4. Also long, meandering, very high rattling, varying in speed and pitch. ☉ NZ 1. R E.NZ

39.4 SOUTH ISLAND (NZPied) **OYSTERCATCHER**[46] *Haematopus finschi* [Huîtrier de Finsch] L 46cm. From very similar pied morph of 39.1 by different pattern of upperwing and back. ✳ Breeds inland on sandbanks and shingle beds in rivers and at lake sides; winters on sandflats and mudflats at coast and in estuaries. ♪ High, sharp, 1- or 2-syllabled *tuweet tuweet*. ☉ NZ 1,2,4,5,10. E.NZ

39.5 BLACK-WINGED STILT *Himantopus himantopus* [Échasse blanche] L 37cm. Polymorph, ♂ ♂ in Br plumage with all-white (**a**) or partly black (**b**) head, this variation also possible in ♀ ♀. ♀ ♀ show brown (not glossy black) backs. N-br ♂ ♂ show pale brown at head and neck and have brown backs. Imm. resembles N-br ♂ but note pale feather tips on upperparts. ✳ Shallow wetlands such as marshes, estuaries, lagoons, paddy fields. ☉ Ha 17,18, NMa, Pa, Mi 4, ?Ma.

39.6 PIED STILT[47] *Himantopus leucocephalus* [Échasse d'Australie] L 36cm. Note black stripe at back of neck reaching nape, missing in Imm. and unlike black-headed morph of 39.5. ✳ Shallow, open wetlands. ♪ Long series of midhigh *whep* notes. ☉ NZ 1–4,9.

39.7 BLACK STILT *Himantopus novaezelandiae* [Échasse noire] L 38cm. Unmistakable. Imm. becomes gradually all-black with age, starting at vent. Shown are also some hybrids (**a, b**) between 39.6 and 39.7. ✳ Gravel- and shingle banks in rivers, shallow ponds and other water bodies. ♪ When disturbed: series of *whep* notes (higher pitched than 39.6). ☉ NZ 1,2. R E.NZ

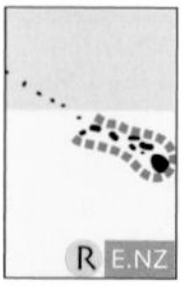

39.8 BLACK-NECKED STILT *Himantopus mexicanus* [Échasse noire] L 38cm. Ssp in the area is *knudseni* (Hawaii Stilt), differing from American mainland ssps by the partly white, not all-black cheeks of ♂. Note the white mark above eyes. Black of upperparts not interrupted at upper mantle. ✳ Ponds, mudflats and wet grassy areas. ♪ High, sharp, sustained *bicbicbic—*. ☉ Ha 1–8. R

39.9 RED-NECKED AVOCET *Recurvirostra novaehollandiae* [Avocette d'Australie] L 44cm. Unmistakable, chestnut head diagnostic. ✳ Prefers shallow inland wetlands. ♪ Long, irregular series of midhigh *wec* notes. ☉ NZ 2. V

40.1 ORIENTAL PRATINCOLE *Glareola maldivarum* [Glaréole orientale] L 29cm. Note tern-like shape with long wings and forked tail, but with different colouring (not white-grey-black) and long legs. ❀ Open grassland, bare fields, coastal mudflats. ♫ Sharp chattering in flock. ☉ Pa, Gu, NMa 1,4, Mi 3,4, Ma, NZ 1–3,10. R

40.2 SHORE PLOVER *Thinornis novaeseelandiae* [Pluvier de Nouvelle-Zélande] L 20cm. Ad. unmistakable. Note compact jizz, long tail projection, bi-coloured bill. ❀ Mainly on rocky shores, boulder-strewn beaches. ♫ Single or vigorous series of *bic* notes. ☉ NZ 1,2,4. E.NZ

40.3 WRYBILL *Anarhynchus frontalis* [Pluvier anarhynque] L 21cm. Unmistakable by bill (curved right), narrow single breast-band, and compact jizz. ❀ Breeds in large areas of shingle and sand in river beds. Winters at coast in shallow estuaries and mudflats. ♫ Very high, thin, well-separated *weet* notes. ☉ NZ 1,2,4. R E.NZ

40.4 MASKED LAPWING[48] *Vanellus miles* [Vanneau soldat] L 34cm. Unmistakable. Note face lappets, wing spurs, red legs. ❀ Open habitats with short grass at shallow wetlands. ♫ Noisy; high, loud, scratchy *kreekreekree* with faster or longer variations. ☉ NZ 1,2,4,5,7,8,10, Fi, Co.

40.5 UPLAND SANDPIPER *Bartramia longicauda* [Maubèche des champs] L 29cm. Long tail, long thin neck and heavy body distinctive. ❀ Normally at open grassland, bare fields, burns. ♫ Slightly descending, rapid chatter or curlew-like, fluted *puréet*. ☉ NZ 1. R

40.6 AMERICAN GOLDEN-PLOVER *Pluvialis dominica* [Pluvier bronzé] L 26cm. Note attenuated jizz. Very similar to 40.7, but in all plumages differing by longer primary projection (normally 4 or 5 primaries project beyond tertials in folded wing). Back in Br plumage with less yellow spots and with flanks fully black (traces of this feature normally still visible in Aug.–Sept., when moulting into N-br plumage, see a). ❀ Short grass, wet fields, lake shores. ♫ Very high, affirmate *awhéet - -*. ☉ NZ 1,2. V

40.7 PACIFIC GOLDEN-PLOVER *Pluvialis fulva* [Pluvier fauve] L 25cm. Cf. 40.6; only 2 or 3 primary tips visible in folded wings. Back in Br plumage with more yellow-gold. Plumage of Imm. on average more yellowish than 40.6. Note moulting plumage (a) with some blackish feathers still visible, mainly on belly. ❀ Mudflats, short grassland, lagoons. ♫ Call: very high *tuweét - -*. ☉ Thr.

40.8 BLACK-BELLIED (or [NZ]Grey) **PLOVER** *Pluvialis squatarola* [Pluvier argenté] L 30cm. Less elegant, more bulky and larger than 40.6–7. No yellow-gold in plumage. Cap less contrasting with eyebrows than 40.6–7. Black axillaries in (N-br) flying bird diagnostic. Note moulting plumage (a). ❀ Mainly coastal on intertidal beaches and mudflats. ♫ High whistled *fluuh* or *flu-weehfluh*. ☉ Thr.

41 DOTTEREL, SANDPLOVER & PLOVERS

41.1 RED-KNEED DOTTEREL *Erythrogonys cinctus* [Pluvier ceinturé]
L 18cm. No similar bird in range. Note faint stripe to flanks of Imm. ❋ Margins at wetlands with tussocks, reeds and rushes. ♪ Mid-high *wrut wrutwrut - -* incorporating short, higher-pitched rattles. ☉ Pa, NZ 1. R

41.2 LESSER SANDPLOVER *Charadrius mongolus* [Pluvier de Mongolie]
L 20cm. Similar in all plumages to 41.3, but with shorter, smaller bill and with shorter, slightly darker legs. ❋ Tidal mudflats, sandy beaches, estuaries. ♪ High, level or slowly rising, fast, musical rattles like *prrrut-prrrut—*. ☉ NMa 1,4, Gu, Pa, Mi, Ma, Fi, Ha, NZ 1,2.

41.3 GREATER SANDPLOVER *Charadrius leschenaultii* [Pluvier de Leschenault] L 24cm. Cf. 41.2. Note heavy bill. ❋ Sandy, shelly and muddy beaches, mudflats, reefs, rock platforms. ♪ Very high, unstructured, shrieking *tirrrweet.* ☉ Gu, NMa 1,4, Pa, Mi, NZ 1,2.

41.4 ORIENTAL PLOVER (or ^{NZ}Dotterel) *Charadrius veredus* [Pluvier oriental] L 24cm. Long yellowish legs diagnostic. ♀ and N-br plumage lack sharply defined breast band. ❋ Mainly inland at dry, sparsely vegetated plains and bare margins of wetlands. ♪ Very high, unstructured, staccato *bic* notes. ☉ NZ 1,2,4,10, Fi 1, Pa. R

41.5 SNOWY (or Kentish) **PLOVER** *Charadrius alexandrinus* [Pluvier à collier interrompu] L 16cm. Very small and pale. Note narrow marks to breast sides, forming an incomplete breast band. ❋ Mainly coastal at beaches of salt and brackish waters. ♪ Series of very high, well-separated, sharp, staccato *bic* notes, sometimes turned into an almost-rattle. ☉ Gu, Mi 4, Pa, NMa 1.

41.6 COMMON RINGED PLOVER *Charadrius hiaticula* [Pluvier grand-gravelot] L 19cm. In all plumages very similar to 41.7 especially in N-br plumage and then not safely separable but by voice. Note in Br plumage black cheeks and white mark above eye. Hawaii is only area where both species 41.6–7 have been recorded together. ❋ Sand and shingle beaches at coast and, less so, inland. Occasionally in farmland. ♪ High *tuwt-tuwt-tuwt—*. ☉ Gu, NMa 1, Pa, Ha, Tu, Ki, Fi, NZ 1. R

41.7 SEMIPALMATED PLOVER *Charadrius semipalmatus* [Pluvier semipalmé] L 18cm. Cf. 41.6. Faint marks above eyes in Br plumage not solid white like those of 41.6. ❋ Beaches and shores free from vegetation; sandbanks, mudflats. ♪ High nasal *tjew-tjew- -*, slurred down to dry rattle. ☉ Ha, Sa 2.

42 DOTTERELS, KILLDEER & PLOVERS

42.1 RED-CAPPED DOTTEREL *Charadrius ruficapillus* [Pluvier à tête rousse] L 15cm. Separable from other plovers by small size, fine bill and bright rufous cap. N-br from very similar 41.5 by lack of white hindneck collar. Note short wing stripe across base of inner primaries. ❋ Prefers bare, open mudflats at lakes. In the past, breeder on river shingle beds. ⊙ NZ 1,2. Irr

42.2 LITTLE RINGED PLOVER *Charadrius dubius* [Pluvier petit-gravelot] L 16cm. Small; appears slender and rather long-legged. Without or with very narrow wing stripe. ❋ Normally avoids the actual seashore; prefers sparsely vegetated flats at shallow inland waters. ♫ Very high, very sharp *picpic-wurow* (last part lower pitched) or almost-rattle *bicbicbic—*. ⊙ Gu, Pa, Mi 4, Ma.

42.3 DOUBLE-BANDED PLOVER[49] (or [NZ]Banded Dotterel) *Charadrius bicinctus* [Pluvier à double collier] L 20cm. Double breast band diagnostic. Imm. like N-br plumage but with buffier eyebrow and lower cheeks. ❋ Breeds on sand, shingle or gravel riverbanks; in winter in variety of habitats such as coastal and inland wetlands, pastures. ♫ Very high, stressed *bic* in loose series. ⊙ NZ 1,2,4,7,10, Fi.

42.4 BLACK-FRONTED DOTTEREL *Elseyornis melanops* [Pluvier à face noire] L 17cm. Unmistakable. Note dark scapulars. ❋ Mainly on gravel riverbeds and nearby farmland. ♫ Very high, scratchy rattles, alternating with sharp, high *bit pip pip bit bit - -*. ⊙ NZ 1,2.

42.5 RED-BREASTED (or [NZ]New Zealand) **DOTTEREL**[50] *Charadrius obscurus* [Pluvier roux] L 27cm. Southern population (**a**) in Br plumage darker than northern population (**b**). Large, thickset jizz with strong, slightly uptilted bill. Nondescript pale N-br plumage. Cf. 40.7. ❋ Ocean beaches, estuarine mudflats, dunes, farmland. On Stewart I. on exposed hilltops when breeding. ♫ Very high, sharp *wrut-wrut—*. ⊙ NZ 1–3. R E.NZ

42.6 KILLDEER *Charadrius vociferus* [Pluvier kildir] L 25cm. Unmistakable. Note rufous rump and long tail. ❋ Open areas, grasslands, fields, marsh. ♫ Loud, shrill *kill-dee*. ⊙ Ha.

42.7 EURASIAN DOTTEREL *Charadrius morinellus* [Pluvier guignard] L 21cm. Br plumage unmistakable. Note in N-br plumage broad, long eyebrow and indistinct white breast bar. ❋ Normally dry plains, but also at coast. ⊙ Ha 18.

43.1 BLACK-TAILED GODWIT[51] *Limosa limosa* [Barge à queue noire]
L 40cm. Note unmarked greyish N-br plumage, which is whiter below than 43.4.
Underwing mainly white. Orange in Br plumage restricted to head, neck and
upper breast. ✳ Winters mainly in estuaries. ♫ High, loud *kipkipkip* or very high,
sharp *wit wit witwit* —. ☉ Gu, NMa 1, Pa, Mi 1,3,4, Ma, Ki, Ha 2,6, NZ 1,2.

43.2 BAR-TAILED GODWIT[52] *Limosa lapponica* [Barge rousse] L 39cm. Note
slightly upturned bill. In N-br plumage slightly streaked above. Shorter-legged
than 43.1. ✳ Winters mainly at or near the sea coast. ♫ Call: *kuwit*. ☉ NMa, Pa,
Mi, Ma, Ki, Fi, Sa, NZ, Ni, Ton, Co, Ha.

43.3 MARBLED GODWIT *Limosa fedoa* [Barge marbrée] L 45cm. Largest
godwit. Buffy tones diagnostic. ✳ Normally at coast. ♫ High resounding,
cackling *tuwuucwuuc*—. ☉ Ha 1,2,6,13.

43.4 HUDSONIAN GODWIT *Limosa haemastica* [Barge hudsonienne]
L 39cm. Black underwing coverts diagnostic. Smaller than 43.1 with darker bill
in N-br plumage. ✳ Coastal waters and marsh. ♫ Very high, hurried, piping
whéetwítwit. ☉ NZ 1,2,4, Ha 6, Ma 1, Fi. R

43.5 FAR EASTERN CURLEW *Numenius madagascariensis* [Courlis de
Sibérie] L 60cm. Very long bill, unstriped face and uniform upperparts
diagnostic. ✳ Mainly coastal, on mudflats, sandflats, estuaries, mangrove.
♫ Melodious, resounding, upswept *uRéet* in series of up to 10. ☉ Pa, Mi 3,4, Fi,
Sa, NMa 1, Gu, Ha 11,13. R

43.6 EURASIAN CURLEW *Numenius arquata* [Courlis cendré] L 55cm. From
43.7**b** by unstriped face sides. From 43.5 by white wedge on back and white belly.
✳ Sea and lake shores, estuaries, sand dunes, riverbanks, grassland. ♫ Beautiful,
mellow, high, loud, slightly trilled *fruuw-fuui-fuuui*. ☉ Ni, NMa 1, Gu. R

43.7 WHIMBREL[53] *Numenius phaeopus* [Courlis corlieu] L 43cm. Ssp *variegatus*
shown (**a**, greyish including barred underwings) and Nom. (**b**, with white wedge
on back; probably not yet seen in the area). Not shown ssp *hudsonicus* (resembles
a, but buffier on underparts). ✳ Mainly coastal. ♫ *bicbicbic*— (up to 7 times).
☉ NMa, Pa, Mi, Ma, Ha, Fi, Sa, Co, NZ.

43.8 BRISTLE-THIGHED CURLEW *Numenius tahitiensis* [Courlis d'Alaska]
L 42cm. Square, pale cinnamon rump diagnostic. ✳ Prefers inland marsh, wet
grassland. ♫ Flight call: melodious *weejowip* or *yourwhéet*. ☉ Ha, Mi, Ma, Ki,
FrPo, Pi, Fi, Ton, Sa, Ni, Co, NMa, Na, NZ 10. R

43.9 LITTLE CURLEW (or [NZ]Whimbrel) *Numenius minutus* [Courlis nain]
L 30cm. From other curlews by small size and short, almost straight bill. Tail
shorter and bill longer than yellow-legged 40.5. ✳ Short grass and bare
cultivation near water bodies. ☉ NZ 1,2, NMa 1, Gu, Pa. V

44 REDSHANKS, GREENSHANKS, SANDPIPERS & YELLOWLEGS

44.1 COMMON REDSHANK *Tringa totanus* [Chevalier gambette] L 28cm. Distinctively red legs and partly red bill. Note broad white trailing edge to wings in flight. ✳ Short grass at edges of wetlands and water bodies. ♫ High, clear *túwee* and high, melodious *tjuutjuutjuu-*. ☉ Gu, Mi 4, Pa, NMa 1.

44.2 NORDMANN'S GREENSHANK *Tringa guttifer* [Chevalier tacheté] L 31cm. Very similar to 44.6, but in N-br plumage with more 2-toned bill, yellower legs and more uniform upperparts. ✳ Normally on/at sandflats, mudbanks, lagoons, mangrove, swamps, meadows, paddy fields. ☉ Gu.

44.3 GREEN SANDPIPER *Tringa ochropus* [Chevalier cul-blanc] L 23cm. Rather compact and short-legged. Note clear-cut breast band. Patterned like 44.9, but more contrastingly black-and-white above; note black underwings (not shown). ✳ Sheltered pools, streams, creeks, sewage ponds. ♫ Loud *weereet-weet-weet*. ☉ Pa, NMa 1.

44.4 SPOTTED REDSHANK *Tringa erythropus* [Chevalier arlequin] L 31cm. N-br plumage like 44.1, but with longer bill and legs. Note that bill is slightly drooped at tip and that red is restricted to basal half of lower mandible. White wedge on back formed like 'cigar'. ✳ Muddy and marshy pools, riverbanks, lagoons, paddy fields. ☉ Mi 3, ?Gu.

44.5 MARSH SANDPIPER *Tringa stagnatilis* [Chevalier stagnatile] L 24cm. From 44.6 by smaller size and straight, thinner bill; from 44.9 by thinner, longer bill and deep white wedge on back. ✳ Muddy and marshy shores of lakes and rivers, streams, pools. ♫ Very high, sharp *chipchip-* (2–5 times) or *tew*. ☉ Pa, Gu, NMa 1, Mi 3,4, NZ 1,2,4, Ha 6,17. R

44.6 COMMON GREENSHANK *Tringa nebularia* [Chevalier aboyeur] L 33cm. Cf. 44.2. Note slightly upturned bill and rather pale head. ✳ Variety of freshwater and saline wetlands, including estuaries, mangrove, sewage ponds. ♫ High, resounding *chewchew-* (2–3 times). ☉ Gu, NMa 1,6, Pa, Mi 3,4, NZ 1,2,4,7. R

44.7 GREATER YELLOWLEGS *Tringa melanoleuca* [Grand Chevalier] L 31cm. Like 44.8 with square white rump and distinctive yellow legs, but bill upturned and longer (longer than length of head). ✳ Mostly on intertidal mudflats and in lagoons. ♫ Very high, rapid, staccato, descending *djípdjipdjip*. ☉ Co, Ha, Ma 11, NMa 4.

44.8 LESSER YELLOWLEGS *Tringa flavipes* [Petit Chevalier] L 24cm. Cf. 44.7, which is larger and slightly more robust. ✳ Coastal and inland wetlands, preferring intertidal flats and lagoons. ♫ Very high, staccato *yipyip-yipyip-yip-yip-yip*. ☉ Ma, Fi 4, FrPo 2, NZ 1,2,4, Ha. R

44.9 WOOD SANDPIPER *Tringa glareola* [Chevalier sylvain] L 21cm. Breast band not clear-cut like 44.3 and bill not so fine as in 44.5. Note rather long legs and distinct eyebrow extending behind eye in N-br plumage. ✳ Secluded parts of marshes, pools, creeks, inundations, sewage ponds. ♫ Very high, staccato, slightly descending *tri-tri-ti-tit*. ☉ Sa 2, Ha 17,18, Ma, Pa, Mi 3,4, NMa.

45 TATTLERS, SANDPIPERS, WILLET & TURNSTONE

45.1 GREY-TAILED (or ^{NZ}Siberian) **TATTLER** *Tringa brevipes* [Chevalier de Sibérie] L 28cm. N-br plumage shows whiter flanks than 45.2; barring below in Br plumage interrupted at central breast and belly. From 45.2 also by calls and paler tail and tail coverts. No white in upperparts. ✳ Rocky coasts, reefs, mudflats, mangrove; may occur at inland wetlands. ♪ High, sharp *tjew* notes, single or in series. ☉ NMa, Pa, Mi, Ma, NZ 1,2,4,10, Ha, Ki, Na, Fi, Tu, Co

45.2 WANDERING TATTLER *Tringa incana* [Chevalier errant] L 28cm. Cf. 45.1. Note dusky flanks in N-br plumage and extensive barring of underparts in Br plumage. ✳ Reefs, rocky coasts. Rarely on mudflats. ♪ High, strong *puweet* or very high *tleet-tleet tleet-tleet-tleet - -* or *wutwutweetweetweet* or descending *bibibi—*. ☉ Thr.

45.3 SOLITARY SANDPIPER *Tringa solitaria* [Chevalier solitaire] L 20cm. Slender and contrastingly dark above. In flight with distinctive tail pattern. ✳ Secluded fresh waters, ponds and creeks. ♪ Very high *swit swit swit*; very high, fluted and rolling *triotriotriotrioh-ukwho-oh*. ☉ Ha 1,6.

45.4 WILLET *Tringa semipalmatus* [Chevalier semipalmé] L 37cm. Note large size, heavy build and strong, straight bill. Wing pattern as in 43.4, but upper tail is barred. Note black underwing coverts, unlike 43.1. ✳ Coastal habitats such as saltmarshes and mudflats. ♪ High, melodious *uk-who-oh who-it-who-it- -*. ☉ Ki 13, Ha. V

45.5 TEREK SANDPIPER *Xenus cinereus* [Chevalier bargette] L 24cm. Unmistakable by long upcurved bill. ✳ Mudflats, estuaries, coral reefs, sandy beaches. ♪ Series of *purrrwéet*. ☉ NMa 1,6, Gu, Pa, Mi 4, Fi 1, Ton, NZ 1,2.

45.6 SPOTTED SANDPIPER *Actitis macularius* [Chevalier grivelé] L 19cm. Like 45.7 with white peak between carpal and breast; tail protruding far beyond folded wing. Br plumage distinctive (note orange, black-tipped bill). N-br plumage not safely separable from N-br 45.7. Imm. with slightly more distinctive patterning above than Ad. In flight with rather shorter white wingbar than 45.7. ✳ Sandy beaches, lagoons, mangroves. ♪ Descending *weetweetweet*; hurried *prrrrWéetprrrWéet*. ☉ Ha, Ma.

45.7 COMMON SANDPIPER *Actitis hypoleucos* [Chevalier guignette] L 20cm. Cf. 45.6. N-br plumage from Br by reduced barring above and wider open collar at breast. ✳ Variety of wetlands, including seashores, estuaries, riverbanks, mangrove, grassland. ♪ Extreme high, sharp *feetfeetfeet*. ☉ NMa, Pa, Mi 3,4, NZ 1,2, Ki, Fi, Sa, Co 1.

45.8 TUAMOTU SANDPIPER *Prosobonia cancellata* [Chevalier des Touamotou] L 16cm. Pale (**a**) and dark (**b**) morph shown, but many intergrades exist. Small with warbler-like bill and long hind toe. Distinctive eyebrow. In flight (not shown) with rounded wings. ✳ Ocean shore, lagoon beaches, bare gravel. ♪ Series of well-separated *beep* and *bic* notes like *beep beep bicbic bic beep - -*. ☉ FrPo 2. R E.FrPo

45.9 RUDDY TURNSTONE *Arenaria interpres* [Tournepierre à collier] L 24cm. Unmistakable in all plumages by colour pattern and short-legged, hunched jizz. ✳ Rocky coasts and nearby open wetlands. ♪ High to very high, irregular, rattling chirps. ☉ Thr.

46 SNIPES & PAINTED-SNIPE

46.1 GREATER PAINTED-SNIPE *Rostratula benghalensis* [Rhynchée peinte] L 26cm. Unmistakable by large, dark eye set in white ring and by pale braces.
❋ Shallow freshwater wetlands with scattered emergent vegetation.
♫ Normally silent; occasionally high, short twitters. ☉ NZ 2. Irr

46.2 SUBANTARCTIC SNIPE[54] *Coenocorypha aucklandica* [Bécassine d'Auckland] L 23cm. Unmistakable. Lacks stripes on back (*Gallinago* snipes have distinct stripes on upperparts). Three ssps known, best identified by range: shown dark *huegeli* (**a**, Snares Is) and Nom. (**b**, Auckland I.); not shown ssp *meinertzhagenae* (Antipodes Is). ❋ Tussock grassland, woodland with dense undergrowth, scrubland. ♫ Like 46.3 but lower-pitched. ☉ NZ 6–8. R E.NZ

46.3 CHATHAM ISLANDS SNIPE *Coenocorypha pusilla* [Bécassine des Chatham] L 19cm. Like 46.2, but in different range. ❋ Bush- and woodland, damp grassy habitats. ♫ Level series of high, sharp whistles *tjew-tjew-tjew—* (10–15 times). ☉ NZ 4. E.NZ

46.4 LATHAM'S (or [NZ]Japanese) **SNIPE** *Gallinago hardwickii* [Bécassine du Japon] L 28cm. Note attenuated jizz. 18 tail feathers. ❋ Coastal wetlands.
♫ Sneezing *snitch*. ☉ NZ 1,2,5, NMa 1, Ma 1. Irr

46.5 COMMON SNIPE *Gallinago gallinago* [Bécassine des marais] L 26cm. In flight with white trailing edge to secondaries diagnostic. 14 tail feathers, the outer ones of 'normal' width. ❋ Marshland, grassy, reedy edges of waterbodies, swampy meadows, estuaries. ♫ When flushed, rises steeply, calling *tweek*.
☉ Mi 4, NMa 1,4, Gu, Ha 2,6,7,18.

46.6 PINTAIL (or [AOU]Pin-tailed) **SNIPE** *Gallinago stenura* [Bécassine à queue pointue] L 26cm. In flight shows pale central wing mirror (formed by central upperwing coverts) and uniform dark underwings. Diagnostic tail structure with 26 feathers. ❋ Marshes. ☉ Ha 18, Mi, NMa 2,4.

46.7 SWINHOE'S SNIPE *Gallinago megala* [Bécassine de Swinhoe] L 28cm. Tail with 20 feathers. ❋ Like 46.5, but in drier situations. ☉ Gu, Mi 3,4, Pa, NMa.

46.8 WILSON'S SNIPE *Gallinago delicata* [Bécassine de Wilson] L 26cm. Main difference with 46.5 are 16 tail feathers. ❋ Grassy edges of shallow ponds. ☉ Ha. *Note:* often considered conspecific with 46.5.

47.1 WILSON'S PHALAROPE *Phalaropus tricolor* [Phalarope de Wilson] L 23cm. Note square white rump and lack of white wingbar in flight; also the needle-thin bill, which is longer than in 47.2, and absence of dark ear coverts in N-br plumage. Legs yellow in N-br and Imm. plumage. ❋ Mainly in inland wetlands, occasionally on sheltered tidal pools, lagoons and estuaries. ♫ In flight short, nasal *tzit* in series. ⊙ NZ 1,2, Ha.

47.2 RED-NECKED PHALAROPE *Phalaropus lobatus* [Phalarope à bec étroit] L 19cm. Note thin bill. Upperparts pattern like in 47.3. ❋ Winters at sea. On migration occasionally on shallow inland waters. ♫ High, sharp *twit* in series. ⊙ Pa, NZ 1,2, Ha 6,7,13, NMa 2. V

47.3 RED (or NZGrey) **PHALAROPE** *Phalaropus fulicarius* [Phalarope à bec large] L 21cm. Rather strong, yellow-based bill. Bulkier in flight than 47.2. ❋ Migrates and winters at sea. ♫ High, strident *zip* in series. ⊙ Ki, NZ 1,2, Ha. R

47.4 ASIAN (or NZAsiatic) **DOWITCHER** *Limnodromus semipalmatus* [Bécassin d'Asie] L 35cm. Dowitchers differ from godwits (Plate 43) by straight, snipe-like, mostly black bill. Darker legs and more distinctly patterned above than 47.5–6. ❋ Lagoons, estuaries, mudflats in sheltered, coastal areas. Occasionally inland. ⊙ NZ 1,2. V

47.5 SHORT-BILLED DOWITCHER *Limnodromus griseus* [Bécassin roux] L 27cm. Not safely separable from 47.6 but by voice. Length of bill of both species varying so not a reliable field mark. ❋ Prefers shallow coastal wetlands, marsh, floodplains. ♫ Call in flight or when flushed: high, rather sharp, rapid *tututu* (2–3 times *tu*). ⊙ Ha.

47.6 LONG-BILLED DOWITCHER *Limnodromus scolopaceus* [Bécassin à long bec] L 27cm. Cf. 47.4–5. ❋ May be found in marsh, at lake margins, floodplains; less in coastal habitats. ♫ Call in flight: very high, whinnying, rapid *wiwiwi* (1–5 times). ⊙ NMa 1, Gu, Ha, Sa 2.

47.7 BUFF-BREASTED SANDPIPER *Tryngites subruficollis* [Bécasseau rousset] L 19cm. Distinctive short, thin bill and buff colour. Note white underwings with dark wrist patch. ❋ Habitats with short grass such as meadows, airfields, pastures, margins of wetlands. ⊙ Ha 6,7,11,17,18, Ma 14, Mi 2,3, ?Sa. V

47.8 RUFF & REEVE *Philomachus pugnax* [Combattant varié] L 29cm (♂, Ruff), 23cm (♀, Reeve). ♂♂ in Br plumage very variable yet easily recognised. N-br ♂ (larger than ♀) with yellow-based bill. Imm. from smaller 47.7 by more elongated head and longer neck. Characteristic narrow dark stripe through rump in flight. ❋ Shallow water, wet grassland, floodplains, marsh. ⊙ NMa, Ma, Ha, Mi 4, Pa, NZ 1,2. R

48.1 BROAD-BILLED SANDPIPER[55] *Limicola falcinellus* [Bécasseau falcinelle] L 17cm. Note long bill with kinked tip. ❋ Sheltered mudflats at coast; occasionaly on reefs, lagoons, swamps, sewage ponds. ♬ Trilling, upslurred *tree'tree'treet*. ☉ NZ 1, Pa.

48.2 RED KNOT[56] *Calidris canutus* [Bécasseau maubèche] L 24cm. Distinctive large size with rather short bill (shorter than similar grey *Tringa* or *Limnodromus* species in N-br plumages). ❋ Coastal mudflats, sandy beaches, saline wetlands near coast. ♬ Silent in winter quarters. ☉ Ha, NZ 1–3,6,7,10, Fi, Sa 2, Gu, Pa.

48.3 GREAT KNOT *Calidris tenuirostris* [Bécasseau de l'Anadyr] L 27cm. Cf. 48.2; differing mainly by size and cleaner white rump. ❋ Coastal mudflats, estuaries, lagoons, ocean beaches; rarely inland. ♬ Silent; incidently in flight soft 2-syllable call. ☉ Gu, Mi 3, Pa, NZ 1,2. R

48.4 SANDERLING *Calidris alba* [Bécasseau sanderling] L 21cm. Normally found at the actual tideline; N-br plumage very pale grey with darker shoulder. ❋ Prefers open sandy beaches at the coast. ♬ Very high, nasal *tweetweetwee*— or sharp *wickwick*. ☉ Thr.

48.5 RED-NECKED STINT *Calidris ruficollis* [Bécasseau à col roux] L 15cm. Most common stint at tropical coasts. Note short bill. N-br plumage not safely separable from 48.7 and 49.5, 49.7 and 49.9. ❋ Coastal mudflats in bays, lagoons and estuaries; occasionally at sandy beaches. ♬ Silent, occasionally soft twittering in feeding flock. ☉ NZ 1,2,4,6, NMa, Pa, Mi 3,4, Ma, Ha, Fi, Sa 2.

48.6 WHITE-RUMPED SANDPIPER *Calidris fuscicollis* [Bécasseau à croupion blanc] L 17cm. White rump and pale base to lower mandible diagnostic. Cf. 49.4 and 49.8 with similar white rump. ❋ Shallow coastal waters, marshes, ponds, lake edges. ♬ Very high, thin *vritvrit vrituhvrit -*.☉ NZ 1,2, Ha 6.

48.7 WESTERN SANDPIPER *Calidris mauri* [Bécasseau d'Alaska] L 16cm. See 48.5, which has shorter bill. Note slightly downcurved bill. ❋ Mudflats, beaches, coastal wetlands such as lake margins and ponds. ♬ Occasionally, thin *cheep*. ☉ NZ 1,2, Ha. R

48.8 LEAST SANDPIPER *Calidris minutilla* [Bécasseau minuscule] L 14cm. Legs paler yellowish green than most other small stints. Upperparts in N-br plumage less solid grey than e.g. 49.7. ❋ Prefers freshwater mudflats with some low vegetation. ♬ Very high, thin *weet weetweet*. ☉ Ha. R

48.9 BAIRD'S SANDPIPER *Calidris bairdii* [Bécasseau de Baird] L 16cm. N-br plumage like 49.8 but with darker legs. Note long wings and closed breast band. ❋ Slightly higher and drier margins and parts and zones of coastal and inland wetlands. ♬ High, slightly rising *peet* (1–2 times) and a trilling *rrrreeeet*. ☉ NZ 1, Ha.

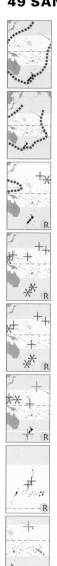

49.1 PECTORAL SANDPIPER *Calidris melanotos* [Bécasseau tacheté]
L 21cm. Note contrast between white belly and tawny coloured, striped breast;
also the pink or greenish bill base and yellowish-green legs. ❋ Drier patches in
freshwater and brackish wetlands, including floodplains and marsh. ♫ Silent,
occasionally *trrrit*. ⊙ Thr. excl. FrPo and Pi. R

49.2 SHARP-TAILED SANDPIPER *Calidris acuminata* [Bécasseau à queue
pointue] L 20cm. From 49.1 by more distinctive cap and broad dark streak
through centre of uppertail coverts. Note striping to lower flanks. ❋ Mudflats
with some grass or weeds. ♫ High, short twitter, combined with sharp *tweetweed*.
⊙ Thr. excl. FrPo 3, Pi. R

49.3 DUNLIN *Calidris alpina* [Bécasseau variable] L 19cm. Rather long, slightly
decurved bill and short primary projection beyond tertials. Many individuals
retain some black belly feathers when the (N) autumn starts before attaining full
N-br plumage. ❋ Estuarine mudflats and other brackish and freshwater wetlands.
♫ Silent, occasionally sharp *treet*. ⊙ Ha, Gu, Pa, Mi, NMa 1. R

49.4 CURLEW SANDPIPER *Calidris ferruginea* [Bécasseau cocorli] L 21cm.
Long decurved bill distinctive. Elegant with rather long neck. ❋ Muddy and
sandy parts of coastal and inland wetlands. ♫ Soft twittering and sharp *chirrip*.
⊙ Ha, Ma 1, Pa, Mi 4, Gu, NZ 1,2,4,6.

49.5 LITTLE STINT *Calidris minuta* [Bécasseau minute] L 13cm. In N-br
plumage not safely separable from other small stints. Note that centre of breast is
white or shows only a faint breast band and that toes are not webbed. ❋ At small
inland waters, coastal mudflats, seashore. ⊙ NMa 1, Pa, Fi, Ha, NZ 1,2.

49.6 LONG-TOED STINT *Calidris subminuta* [Bécasseau à longs doigts]
L 15cm. Note rather upright stance with long neck, yellowish legs, rather sharp
division of darker breast from white belly and pale base to lower mandible.
❋ Mainly in inland wetlands and at coastal and shallow water bodies. ⊙ Pa, Mi
3,4, ?Gu, Ha 17, Ki 13, NMa 1, NZ 2.

49.7 TEMMINCK'S STINT *Calidris temminckii* [Bécasseau de Temminck]
L 14cm. Tail shows all-white outer feathers in flight and projects beyond
wingtips at rest. ❋ Inland wetlands with dense vegetation; unlikely to be seen on
coastal mudflats and lagoons. ⊙ Gu, NMa 1.

49.8 STILT SANDPIPER *Calidris himantopus* [Bécasseau à échasses] L 21cm.
Note elongated body and long legs. ❋ Shallow freshwater wetlands, rarely along
coast. ♫ Thin *trrp*. ⊙ NZ 2, Ha 6. V

49.9 SEMIPALMATED SANDPIPER *Calidris pusilla* [Bécasseau semipalmé]
L 14cm. Very difficult to separate from other small stints. Note absence of
complete breast band in N-br plumage. Toes partially webbed. ❋ Sandy beaches,
shallow lagoons, tidal flats, inland wetlands. ♫ Very high *tjerrup trrup jupjupjup*.
⊙ Ha, ?Ma, ?NZ.

50 SKUAS & JAEGERS

50.1 BROWN (or [NZ]Southern) **SKUA**[57] *Stercorarius antarcticus* [Labbe antarctique] L 58cm. From 50.3–5 by broad, blunt-pointed wings and stocky body. Note dark brown plumage with variable pale streaking on nape and mantle. No or very little white in feathers at base of upper mandible. ❋ Nests on short grass between tussocks, sedge or tall grass clumps or on gravel or bare rock, often near penguin or petrel colonies. After breeding disperses offshore, occasionally inshore. ♫ High, gull-like *tuw-tuw*—. ☉ Breeds NZ 4–8.

50.2 SOUTH POLAR SKUA *Stercorarius maccormicki* [Labbe de McCormick] L 53cm. From 50.3–5 by broad, blunt-tipped wings and stocky body. Pale morph (**a**) shows marked contrast between body and wings. Most dark-morph individuals show some white at base of upper mandible (**b**). Juveniles are basically grey, not brown. ❋ Normally off coast. ♫ Low, squeaky shrieks *wew-wew*—. ☉ Most numerous near NZ in Dec.–Feb. R

50.3 PARASITIC JAEGER (or [NZ]Arctic Skua) *Stercorarius parasiticus* [Labbe parasite] L 43cm, excl. streamers of up to 11cm. Note pointed wingtips. Polymorph, from all-dark (**a**) to pale (**b**) and all types of intermediates (**c**, head of example); Juvs also polymorph, but basically rusty coloured. Cf. very similar 50.4, which is heavier like a large gull. ❋ Mostly inshore incl. estuaries. ♫ Alarm call: mewing *tjew* or *tuweh*. ☉ Most numerous near NZ in Nov.–Feb.

50.4 POMARINE JAEGER (or [NZ]Skua) *Stercorarius pomarinus* [Labbe pomarin] L 49cm, excl. streamers of up to 11cm. From 50.3 by heavier build, slightly darker plumage, coarse, not soft gradients between dark and white plumage parts, lack of white at bill base; note also the difference in bill colouring (cf. heads). Imms from 50.3 by different basic colouring and by pattern on wrists at undersides (see markings on plate, indicating distinct (50.4) or vague (50.3) second pale arc in wing undersides as shown). Blunt streamers diagnostic, but often missing. ❋ Mainly marine. ♫ High to very high squeaky *wewewewew*—. ☉ Present near NZ in Oct.–Mar.

50.5 LONG-TAILED JAEGER (or [NZ]Skua) *Stercorarius longicaudus* [Labbe à longue queue] L 37cm, excl streamers of up to 22cm. Ad. lacks pale underwing patch; note 2 or 3 white primary shafts on upperside and striking contrast between black secondaries and pale upperwing coverts. Only pale Ad. morphs exist; Imms very variable: dark (**a**) and pale (**b**) morph shown, best separable from heavier Imms 50.3–4 by longer central tail feathers and less marked pale patch at base of primaries on underwing. ❋ Offshore and open ocean. ♫ High *trew-trew-trew*—. ☉ May reach NZ 1. R

51.1 RING-BILLED GULL *Larus delawarensis* [Goéland à bec cerclé] L 50cm. From smaller 51.10 by diagnostic, distinctly banded bill, yellow eyes and smaller mirror in wings. 1st W from 1st W 51.10 by paler grey saddle and wing panel and by some markings along scapular (grey in 51.10). 2nd W like Ad. but with more black on outer wing. ❊ Coast, lakes, rubbish dumps, wet meadows. ♫ Very high, angry *píow píow - -*. ☉ Ki 3, Ha. R

51.2 BLACK-TAILED GULL *Larus crassirostris* [Goéland à queue noire] L 46cm. Note distinct, red-tipped bill, rather dark upperparts and lack of white mirrors in wings. Note also grey saddle and wingbar of generally darkish brown 2nd W. ❊ Could occur on the coast, incl. bays and harbours. ☉ Gu.

51.3 RED-BILLED GULL[58] *Larus scopulinus* [Mouette scopuline] L 37cm. In all plumages from 51.4 mainly by range. ❊ Inshore, beaches, estuaries, urban areas, developed country, fields, wetlands. Occasionally in highlands. ♫ High, hoarse screams *wraaah - -*. ☉ NZ 1–7. E.NZ

51.4 SILVER GULL *Larus novaehollandiae* [Mouette argentée] L 41cm. Like 51.3 (which see) with red bill and legs and with yellow eyes, but differing by range. ❊ Beaches, shores, inland fields, rubbish dumps. ☉ NZ 10, ?FrPo, Fi.

51.5 BLACK-BILLED GULL *Larus bulleri* [Mouette de Buller] L 37cm. Unmistakable. No black or brown markings to head. Reduced black to wingtips. ❊ Breeds on shingle or gravel riverbanks; outside Br season found along the coast, agricultural land, wet pastures. ♫ Hoarse screams (higher pitched than 51.3). ☉ NZ 1–3,5. R E.NZ

51.6 BLACK-HEADED GULL *Larus ridibundus* [Mouette rieuse] L 40cm. Most distinct feature is white leading edge to outer wings. Bill shape, bill colour and/or white eye crescents differ from those of 51.7–9. ❊ Coastal and inland wetlands, meadows, pasture land, refuse dumps. ♫ High, drawn-out *sreeuw sreeuw*. ☉ Mi 4, Pa, NMa, Ha.

51.7 FRANKLIN'S GULL *Larus pipixcan* [Mouette de Franklin] L 35cm. See 51.6. Note dark upperparts and blackish face mask in all plumages. ❊ Coastal and inland waters, marsh, grassland, refuse dumps. ♫ Very high, sharp *wekwek—*. ☉ Ha, Ma, Mi 2,3, Ki, FrPo 3, NZ 1,2,10. R

51.8 BONAPARTE'S GULL *Larus philadelphia* [Mouette de Bonaparte] L 29cm. From 51.6 by black, not brown hood, black bill and different underwing pattern. ❊ Normally in coastal areas. ♫ Low frog-like bickering. ☉ Ha.

51.9 LAUGHING GULL *Larus atricilla* [Mouette atricille] L 43cm. Note long, slightly downcurved bill and black leading edge of wings. ❊ Dumps, sewage ponds, beaches, harbours. Strictly coastal. ♫ Very high *mèew mèew mèew*. ☉ Ha, Ma 3, Ki, Mi 3, Fi 1,3, Sa 2, Co, FrPo 3, NZ 1,2,10. R

51.10 MEW GULL *Larus canus* [Goéland cendré] L 43cm. Text and map opposite plate 52.

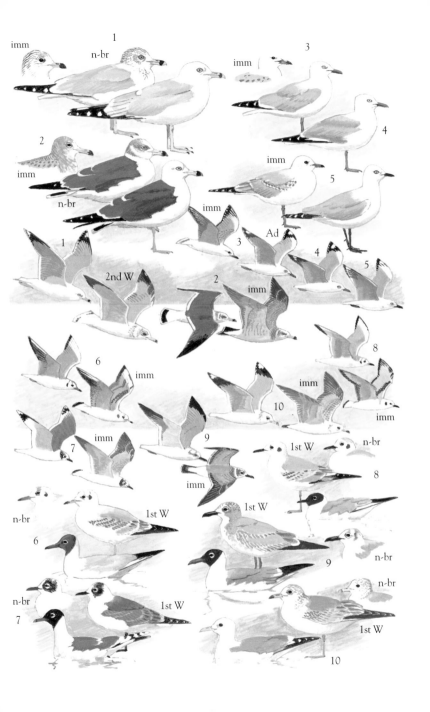

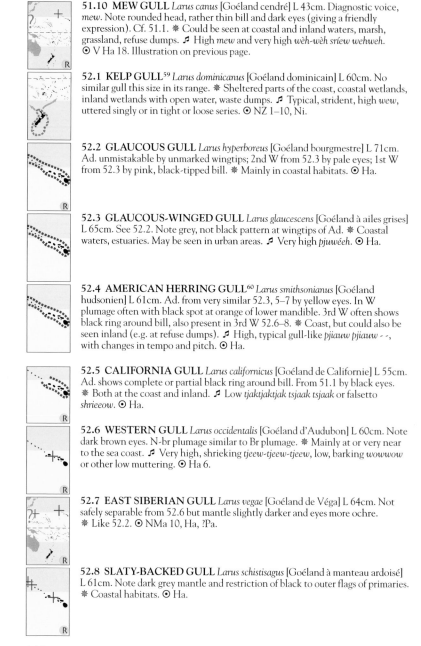

51.10 MEW GULL *Larus canus* [Goéland cendré] L 43cm. Diagnostic voice, *mew*. Note rounded head, rather thin bill and dark eyes (giving a friendly expression). Cf. 51.1. ❋ Could be seen at coastal and inland waters, marsh, grassland, refuse dumps. ♫ High *mew* and very high *wèh-wèh srìew wehweh*. ☉ V Ha 18. Illustration on previous page.

52.1 KELP GULL[59] *Larus dominicanus* [Goéland dominicain] L 60cm. No similar gull this size in its range. ❋ Sheltered parts of the coast, coastal wetlands, inland wetlands with open water, waste dumps. ♫ Typical, strident, high *wew*, uttered singly or in tight or loose series. ☉ NZ 1–10, Ni.

52.2 GLAUCOUS GULL *Larus hyperboreus* [Goéland bourgmestre] L 71cm. Ad. unmistakable by unmarked wingtips; 2nd W from 52.3 by pale eyes; 1st W from 52.3 by pink, black-tipped bill. ❋ Mainly in coastal habitats. ☉ Ha.

52.3 GLAUCOUS-WINGED GULL *Larus glaucescens* [Goéland à ailes grises] L 65cm. See 52.2. Note grey, not black pattern at wingtips of Ad. ❋ Coastal waters, estuaries. May be seen in urban areas. ♫ Very high *pjuwéeh*. ☉ Ha.

52.4 AMERICAN HERRING GULL[60] *Larus smithsonianus* [Goéland hudsonien] L 61cm. Ad. from very similar 52.3, 5–7 by yellow eyes. In W plumage often with black spot at orange of lower mandible. 3rd W often shows black ring around bill, also present in 3rd W 52.6–8. ❋ Coast, but could also be seen inland (e.g. at refuse dumps). ♫ High, typical gull-like *pjiauw pjiauw - -*, with changes in tempo and pitch. ☉ Ha.

52.5 CALIFORNIA GULL *Larus californicus* [Goéland de Californie] L 55cm. Ad. shows complete or partial black ring around bill. From 51.1 by black eyes. ❋ Both at the coast and inland. ♫ Low *tjaktjaktjak tsjaak tsjaak* or falsetto *shrieeow*. ☉ Ha.

52.6 WESTERN GULL *Larus occidentalis* [Goéland d'Audubon] L 60cm. Note dark brown eyes. N-br plumage similar to Br plumage. ❋ Mainly at or very near to the sea coast. ♫ Very high, shrieking *tjeew-tjeew-tjeew*, low, barking *wowwow* or other low muttering. ☉ Ha 6.

52.7 EAST SIBERIAN GULL *Larus vegae* [Goéland de Véga] L 64cm. Not safely separable from 52.6 but mantle slightly darker and eyes more ochre. ❋ Like 52.2. ☉ NMa 10, Ha, ?Pa.

52.8 SLATY-BACKED GULL *Larus schistisagus* [Goéland à manteau ardoisé] L 61cm. Note dark grey mantle and restriction of black to outer flags of primaries. ❋ Coastal habitats. ☉ Ha.

1st W

2nd W

2nd W

2nd W

2nd W

2nd W

2nd W

2nd W

n-br

1

3rd W

n-br

2

1st W

imm

3

3rd W

4

5

6

7

8

53 GULLS, KITTIWAKE & TERNS

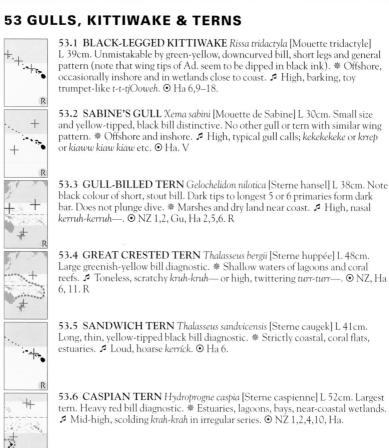

53.1 BLACK-LEGGED KITTIWAKE *Rissa tridactyla* [Mouette tridactyle] L 39cm. Unmistakable by green-yellow, downcurved bill, short legs and general pattern (note that wing tips of Ad. seem to be dipped in black ink). ✳ Offshore, occasionally inshore and in wetlands close to coast. ♫ High, barking, toy trumpet-like *t-t-tjOoweh*. ☉ Ha 6,9–18.

53.2 SABINE'S GULL *Xema sabini* [Mouette de Sabine] L 30cm. Small size and yellow-tipped, black bill distinctive. No other gull or tern with similar wing pattern. ✳ Offshore and inshore. ♫ High, typical gull calls; *kekekekeke* or *krrep* or *kiaww kiaw kiaw* etc. ☉ Ha. V

53.3 GULL-BILLED TERN *Gelochelidon nilotica* [Sterne hansel] L 38cm. Note black colour of short, stout bill. Dark tips to longest 5 or 6 primaries form dark bar. Does not plunge dive. ✳ Marshes and dry land near coast. ♫ High, nasal *kerruh-kerruh—*. ☉ NZ 1,2, Gu, Ha 2,5,6. R

53.4 GREAT CRESTED TERN *Thalasseus bergii* [Sterne huppée] L 48cm. Large greenish-yellow bill diagnostic. ✳ Shallow waters of lagoons and coral reefs. ♫ Toneless, scratchy *kruh-kruh—* or high, twittering *turr-turr—*. ☉ NZ 6, 11. R

53.5 SANDWICH TERN *Thalasseus sandvicensis* [Sterne caugek] L 41cm. Long, thin, yellow-tipped black bill diagnostic. ✳ Strictly coastal, coral flats, estuaries. ♫ Loud, hoarse *kerríck*. ☉ Ha 6.

53.6 CASPIAN TERN *Hydroprogne caspia* [Sterne caspienne] L 52cm. Largest tern. Heavy red bill diagnostic. ✳ Estuaries, lagoons, bays, near-coastal wetlands. ♫ Mid-high, scolding *krah-krah* in irregular series. ☉ NZ 1,2,4,10, Ha.

53.7 GREY-BACKED TERN *Onychoprion lunata* [Sterne à dos gris] L 38cm. Very similar to 53.8, but slightly greyer above; flight feathers darker than coverts, especially showing when perched. ✳ Breeds on bare ground of islands. Disperses offshore after breeding. ♫ Gull-like shrieks. ☉ Thr. excl. NZ, where V.

53.8 BRIDLED TERN *Onychoprion anaethetus* [Sterne bridée] L 37cm. Cf. 53.7. Less dark above than 53.9 with narrower eyebrow. ✳ Inshore and offshore. ♫ Nasal, staccato *wrep-wrep—*. ☉ Ton, WaF, Fi, Sa, NZ 2, Pa, Ma. R

53.9 SOOTY TERN *Onychoprion fuscata* [Sterne fuligineuse] L 41cm. ✳ Breeds on open rock and coral or sand near coast. Disperses after breeding to inshore and offshore waters. ♫ Croaking, sometimes shrill *wrah-wrah* or very high, nasal twittering. ☉ Thr. excl. NZ, where V. R

54 TERNS

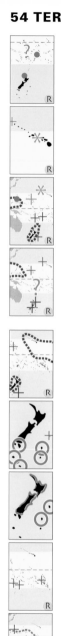

54.1 FAIRY TERN[61] *Sternula nereis* [Sterne néréis] L 25cm. Br plumage from 54.2–3 by cut-off white above eyes and lack of black point to bill. Note bill pattern of N-br plumage. ✹ Coastal habitats such as reefs, estuaries, bays. ♪ Slightly goose-like snarls and very high, slightly descending, nasal, rapid *wiederweet*. ☉ Sa 2, ?Fi, NZ 1.

54.2 LEAST TERN *Sternula antillarum* [Petite Terne] L 23cm. Only seen at Hawaii Is. Not safely separable from 54.3, but note that rump and tail are grey, not white like 54.3. ✹ Wetlands close to the sea coast. ♪ Very high, drawn-up *wriít*. ☉ Ha 6,11,17.

54.3 LITTLE TERN[62] *Sternula albifrons* [Sterne naine] L 25cm. Rump white. Cf. 54.2. ✹ Coastal wetlands; also ocean beach. ♪ Very high, raspy *tweet -*. ☉ Ha 10–18, NMa 1, Gu, Pa, Mi, Ma, NZ 1–6,9,11, Fi 13, Sa 2.

54.4 COMMON TERN[63] *Sterna hirundo* [Sterne pierregarin] L 36cm. Ssp *longipennis* shown (**a**, with black bill and legs) and Nom. (**b**, with red, black-pointed bill and red legs). Note rather long bill, darkish edge to upperwings and tail length (not protruding beyond folded wings). ✹ Offshore, inshore and occasionally in near-coastal wetlands. ♪ Very high *kree-eh* or *kreehkreeh—*. ☉ Gu, NMa 1, Pa, Mi 2,3,4, Ma, Co, Ha, Fi?, NZ 1,2. R

54.5 ARCTIC TERN *Sterna paradisaea* [Sterne arctique] L 35cm. Note very short legs. Longer-tailed and shorter-billed than 54.4. Blackish tips to primaries form narrower dark bar on underwings than 54.4. ✹ Offshore; also in estuaries and reefs. ♪ High *eéeehr*. ☉ Ma, NZ, Ha. R

54.6 ANTARCTIC TERN[64] *Sterna vittata* [Sterne couronnée] L 38cm. Seen in Br plumage when 54.4–5 are in N-br plumage. Note short bill. Bar along tips of primaries on underside of wings is rather pale. ✹ Coastal and inland wetlands, incl. estuaries, lagoons, swamps, lakes rivers, sewage ponds. ♪ Varied; mid-high, scratchy *wruhwruh-* or *k'wruh-k'wruh-*. ☉ NZ 3–9.

54.7 WHITE-FRONTED TERN *Sterna striata* [Sterne tara] L 30cm. Note pale upperparts, narrow black line along leading edge of wings, long bill and dark red legs. ✹ Nests mainly on shingle riverbanks, often at higher altitudes, but can also be seen over pastures, fields, lakes, farm dams. After breeding diperses to sheltered coastal habitats. ♪ Mid-high *tzit tzit - -*. ☉ NZ 1–6.

54.8 ROSEATE TERN *Sterna dougallii* [Sterne de Dougall] L 38cm. Bill only red at start of breeding season, rapidly becoming all-black after hatching of eggs. Very pale above; tail trailers long. Note dark wedge along leading edge of wings. Faint darker bar along edges of primaries at underwings. ✹ Mainly offshore. ♪ Very high *weewee*. ☉ Ton, FrPo 2, Fi. R

54.9 BLACK-NAPED TERN *Sterna sumatrana* [Sterne diamant] L 35cm. Very pale, almost all white. Tail protruding slightly beyond folded wings. Note black line along leading edge of first primary. ✹ At and around coral reefs and rocky islands. ♪ Very high *tjeeptjeep tjeerp-tjeerp*. ☉ Thr. excl. Ha and NZ.

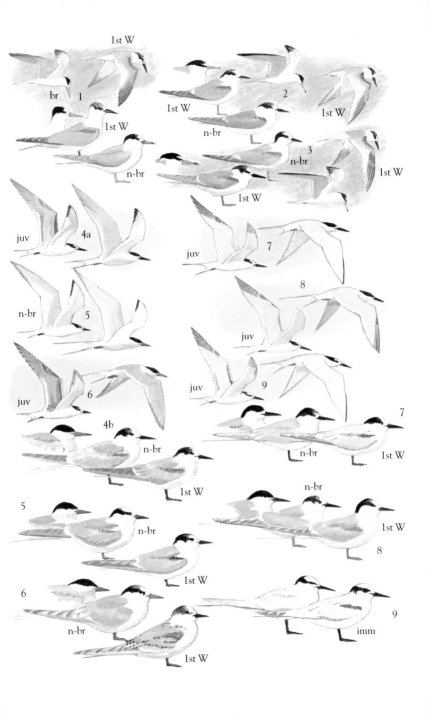

1st W

br 1

1st W

n-br

1st W

n-br

3
n-br

1st W

juv 4a

n-br 5

juv 6

4b
n-br
1st W

5
n-br
1st W

6
n-br
1st W

juv 7

8

juv

juv 9

n-br
1st W 7

n-br
1st W 8

9
imm

55.1 WHITE-WINGED ([NZ]Black) **TERN** *Chlidonias leucopterus* [Guifette leucoptère] L 25cm. Br unmistakable by black underwing coverts; N-br and 1st W from 55.2 by lack of blackish patch at sides of upper breast. ✱ Coastal and inland wetlands, incl. estuaries, harbours, lagoons, swamps, lakes, rivers, sewage ponds. ♪ Silent, occasionally high *kritkrit-*. ☉ Na, NZ 1,2, Pa, Mi 4, Gu, NMa 1,2. R

55.2 BLACK TERN *Chlidonias niger* [Guifette noire] L 25cm. Cf. 55.1. Note white underwing coverts in Br plumage. ✱ On migration may occur in any type of wetland. ♪ High, scratchy *shrets shret - -*, accelerated to *–sretsretsret*. ☉ Ha.

55.3 WHISKERED TERN *Chlidonias hybrida* [Guifette moustac] L 26cm. Br from larger 55.4 by darker belly and dark red, not orange bill. Note uniform, very pale upperparts of N-br plumage. Saddle of Imm. rather orange. ✱ Shallow freshwater wetlands. ♪ Silent, occasionally frog-like *kreek*. ☉ ?Ha, NZ 1,2, Pa, Mi 4, Gu, NMa 1. R

55.4 BLACK-FRONTED TERN *Chlidonias albostriatus* [Guifette des galets] L 30cm. Note striking orange-yellow bill and legs, white cheek streak and white rump. ✱ Breeds at higher altitudes on shingle riverbanks, occasionally also on lakeshores. Forages in nearby agricultural areas. After breeding, mainly in sheltered coastal habitats. ♪ High twittering *kiurrr* or *t't't'tirr*. ☉ NZ 1–3,5. R E.NZ

55.5 GREY NODDY[65] (or [NZ]Ternlet) *Procelsterna albivitta* [Noddi gris] L 30cm. From smaller 55.6 by white head and breast. ✱ Inshore and offshore off rocky islands. ♪ Upslurred rattling *purrr - -*. ☉ Thr. excl. Ha 1–8 and Ki 2.

55.6 BLUE (-[NZ]grey) **NODDY**[66] *Procelsterna cerulea* [Noddi bleu] L 27cm. Cf. 55.5. ✱ Lagoons and inshore waters of rocky islands. ♪ Drawn-out, slightly upslurred, mewing *sreeeeh*. ☉ Ha 13–18, Ma, Ki, Tu, Sa, Fi, Ton, FrPo.

55.7 WHITE TERN *Gygis alba* [Gygis blanche] L 28cm. 4 ssps are involved: Nom. (**a**, all-black bill, W and NW Pacific incl. Hawaii), *candida* (**b**, with blue-based bill and dark primary shafts, SC Pacific incl. FrPo 3), *leucopes* (not shown, like **a**, but often with white legs, SE Pacific) and smaller *microrhyncha* (not shown, with thin, all-black bill, FrPo 3 and Ki; often treated as species). ✱ Coral islands, with trees and bushes. ♪ High, dry, rhythmic shrieks. ☉ Thr. excl. Ha 1–8. R

55.8 BROWN NODDY *Anous stolidus* [Noddi brun] L 42cm. From smaller 55.9 by less uniform plumage, longer white cap, relatively shorter bill and, in flight, pale wingbars. ✱ Inshore and offshore. ♪ Low gurgling and croaks in colony. ☉ Thr. excl. SI. R

55.9 BLACK NODDY *Anous minutus* [Noddi noir] L 37cm. Note uniform dark plumage, short white cap and long, thin bill. ✱ Inshore and offshore. ♪ Mid-high bleating and dry rattles. ☉ Thr.

56 MURRELETS, AUKLETS, PUFFINS, SANDGROUSE & PIGEON

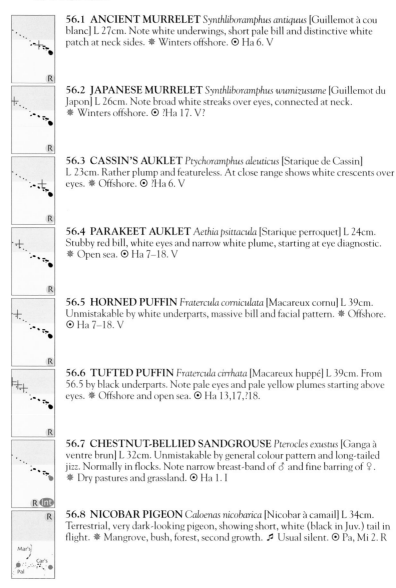

56.1 ANCIENT MURRELET *Synthliboramphus antiquus* [Guillemot à cou blanc] L 27cm. Note white underwings, short pale bill and distinctive white patch at neck sides. ❋ Winters offshore. ☉ Ha 6. V

56.2 JAPANESE MURRELET *Synthliboramphus wumizusume* [Guillemot du Japon] L 26cm. Note broad white streaks over eyes, connected at neck. ❋ Winters offshore. ☉ ?Ha 17. V?

56.3 CASSIN'S AUKLET *Ptychoramphus aleuticus* [Starique de Cassin] L 23cm. Rather plump and featureless. At close range shows white crescents over eyes. ❋ Offshore. ☉ ?Ha 6. V

56.4 PARAKEET AUKLET *Aethia psittacula* [Starique perroquet] L 24cm. Stubby red bill, white eyes and narrow white plume, starting at eye diagnostic. ❋ Open sea. ☉ Ha 7–18. V

56.5 HORNED PUFFIN *Fratercula corniculata* [Macareux cornu] L 39cm. Unmistakable by white underparts, massive bill and facial pattern. ❋ Offshore. ☉ Ha 7–18. V

56.6 TUFTED PUFFIN *Fratercula cirrhata* [Macareux huppé] L 39cm. From 56.5 by black underparts. Note pale eyes and pale yellow plumes starting above eyes. ❋ Offshore and open sea. ☉ Ha 13,17,?18.

56.7 CHESTNUT-BELLIED SANDGROUSE *Pterocles exustus* [Ganga à ventre brun] L 32cm. Unmistakable by general colour pattern and long-tailed jizz. Normally in flocks. Note narrow breast-band of ♂ and fine barring of ♀. ❋ Dry pastures and grassland. ☉ Ha 1. I

56.8 NICOBAR PIGEON *Caloenas nicobarica* [Nicobar à camail] L 34cm. Terrestrial, very dark-looking pigeon, showing short, white (black in Juv.) tail in flight. ❋ Mangrove, bush, forest, second growth. ♫ Usual silent. ☉ Pa, Mi 2. R

57 PIGEON & GROUND-DOVES

57.1 METALLIC PIGEON *Columba vitiensis* [Pigeon à gorge blanche] L 39cm. Nom. (**a**) shown and ssp *castaneiceps* (**b**, darker and more blue-green below). Note distinctive white throat, colour pattern of bill and unbanded tail. ✳ Forest, second growth. ♫ Low, resounding *woooh woooh woooh - -*. ☉ Fi, Sa.

57.2 TOOTH-BILLED PIGEON *Didunculus strigirostris* [Diduncule strigirostre] L 35cm. Unmistakable by bill shape. ✳ Mature forest. ☉ Sa. R E.Sa

57.3 FERAL PIGEON *Columba livia* [Pigeon biset] L 33cm. Because they descend from the wild old-world Rock Pigeon most feral pigeons share ancestral features such as white rump, wingbars, dark terminal tail bar and green neck colouring; some are completely or partly and irregular white, black and/or reddish (**a**). ✳ Urban areas; occasionally at cliffs away from settlement. Unlikely to be found in forest. ☉ Thr. I

57.4 CAROLINE ISLANDS GROUND-DOVE *Gallicolumba kubaryi* [Gallicolombe de Kubary] L 28cm. From 57.5 mainly by range. ♂ and ♀ similar. ✳ Ravines with thickets, plantations and in natural forest. ☉ Mi 2,3. R E.Mi

57.5 POLYNESIAN GROUND-DOVE *Gallicolumba erythroptera* [Gallicolombe érythroptère] L 25cm. Cf. 57.4. Some ♂♂ have all-white heads (**a**). ♀ much paler than ♂ and with tawny underparts. ✳ Littoral scrub of some inhabited islands. ☉ FrPo 5–8. R E.FrPo

57.6 FRIENDLY GROUND-DOVE *Gallicolumba stairi* [Gallicolombe de Stair] L 26cm. Note range. ♀♀ occur in 2 colour morphs; some resemble ♂, others lack white at breast shield (**a**); both morphs found in Fiji, while in Tonga and Samoa most ♀♀ resemble ♂♂. ✳ Forest, bamboo thickets. ☉ Sa, ?Ki, WaF 2, Ton 5,8,9, Fi, Ma. R

57.7 WHITE-THROATED GROUND-DOVE *Gallicolumba xanthonura* [Gallicolombe pampusane] L 26cm. Unmistakable in range if seen well. Rather arboreal. ✳ Forest canopy, but on Yap feeds largely on the ground. ☉ Mi 4, NMa 1–8,12, Gu.

57.8 MARQUESAS GROUND-DOVE *Gallicolumba rubescens* [Gallicolombe des Marquises] L 20cm. Variable; especially width and length of white band in wings and tail may be longer or narrower than shown; also head may be darker or almost white. ✳ Shrubs. ☉ FrPo 9. R E.FrPo

57.9 PALAU GROUND-DOVE *Gallicolumba canifrons* [Gallicolombe de Palau] L 22cm. No similar bird in Palau. ✳ Forest. ☉ Pa 3–6. R E.Pa

58 PIGEONS

58.1 NEW ZEALAND PIGEON[67] *Hemiphaga novaeseelandiae* [Carpophage de Nouvelle-Zélande] L 48cm. Shown are Nom. (**a**) and paler ssp *chathamensis* (**b**). Unmistakable by large size, sharp demarcation white–dark green on breast and noisy swish in flight. ✳ Prefers native forest up to 1,100m; also in parks, gardens, plantations. ♫ Sudden, descending *óoooh*. ☉ NZ 1–3. R E.NZ

58.2 MICRONESIAN IMPERIAL-PIGEON *Ducula oceanica* [Carpophage de Micronésie] L 44cm. From 58.3 by range and rufous belly. ✳ Canopy of forest and plantations. Also in mangrove. ☉ Mi 1–4, Pa 1,2,6, Na, Ma 5,10–12,19, ?Ki. R

58.3 PACIFIC IMPERIAL-PIGEON *Ducula pacifica* [Carpophage pacifique] L 39cm. Cf. 58.2. From 58.6 by enlarged cere, contrasting back, gloss to mantle and wings. ✳ Forest canopy, littoral scrub. ♫ Slightly lowered *rrrooooh* or low *oohóoroh*. ☉ Tu, Ki 2, WaF, Tok, Ton, Co 1,3,4,9,10,11, Sa, Ni, Fi (excl. large islands).

58.4 POLYNESIAN IMPERIAL-PIGEON *Ducula aurorae* [Carpophage de la Société] L 51cm. Unmistakable in range by large size. Note reddish eyes. ✳ Dense forest. ☉ FrPo 4,11. R E.FrPo

58.5 MARQUESAS (or Nukuhira) **IMPERIAL-PIGEON** *Ducula galeata* [Carpophage des Marquises] L 55cm. Not in range of 58.4. Note shape of enlarged cere. ✳ Forest above 700m. ☉ FrPo 12,13. R E.FrPo

58.6 PEALE'S (or Barking) **IMPERIAL-PIGEON** *Ducula latrans* [Carpophage de Peale] L 41cm. Cf. 58.3. Cere not enlarged. Shows tawny vent and chestnut underwing (not slate-grey like 58.3). ✳ Mature forest up to 1,000m. ♫ Characteristic, short barks *wuh wuh* in irregular series. ☉ Fi, all main islands. E.Fi

59 FRUIT-DOVES

59.1 CRIMSON-CROWNED FRUIT-DOVE *Ptilinopus porphyraceus* [Ptilope de Clémentine] L 23cm. 3 ssps: Nom. (**a**, no purple in belly band and terminal tail-bar greyish, Fiji), *ponapensis* (**b**, like **a**, but with indistinct belly band and yellow terminal tail-bar, Chuuk and Pohnpei) and *fasciatus* (**c**, like **a**, but with partly purple belly patch, Samoa). From Samoa ssp of 59.9 by yellow-orange vent; from Fiji ssp of 59.9 by restriction to small islands. ❋ Forest, mangrove, scrub. ☉ Sa, A.Sa 4, WaF, Ton, Fi (small islands), Ni, Mi 2,3,4.

59.2 COOK ISLANDS FRUIT-DOVE *Ptilinopus rarotongensis* [Ptilope de Rarotonga] L 22cm. Nom. (**a**, with purplish breast patch) shown and ssp *goodwini* (**b**, with green breast band). No similar fruit-dove on Cook. ❋ Dense forest (Rarotonga) or wooded habitats incl. plantations (elsewhere). ♫ Mid-high accelarating and slightly descending *cooh-cooh*—. ☉ Co 1,3. R E.Co

59.3 GREY-GREEN FRUIT-DOVE *Ptilinopus purpuratus* [Ptilope de la Société] L 20cm. No similar fruit-dove on Society Is. ❋ Forest, plantations. ♫ Drawn-out, slightly rising, indignant, accelerating series *wooow wooow-wooow- -* (9–10 times). ☉ FrPo 4,14–19. E.FrPo

59.4 MAKATEA FRUIT-DOVE *Ptilinopus chalcurus* [Ptilope de Makatéa] L 22cm. No similar fruit-dove on Makatea I. ❋ Wooded areas of the island. ☉ FrPo 11. R E.FrPo

59.5 ATOLL FRUIT-DOVE *Ptilinopus coralensis* [Ptilope des Touamotou] L 23cm. No similar fruit-dove in the Tuamotu Archipelago. ❋ Woodland, scrub, overgrown plantations. ☉ FrPo 2, excl. 11. R E.FrPo

59.6 RAPA FRUIT-DOVE *Ptilinopus huttoni* [Ptilope de Hutton] L 31cm. No similar fruit-dove on Rapa I. ❋ Dense forest. ♫ Rather long series of descending, accelerating *wuh* notes, preceded by *ooh-rurr* (*rurr* low). ☉ FrPo 29 (Rapa). R E.FrPo

59.7 WHITE-CAPPED FRUIT-DOVE *Ptilinopus dupetithouarsii* [Ptilope de Petit Thouars] L 20cm. Nom. (**a**) shown and ssp *viridior* (**b**, with smaller orange-yellow breast patch). White crown diagnostic. ❋ Forest. ♫ Mid-high, slightly downslurred *oowooh*. ☉ FrPo 12,13,20,23,25,26. E.FrPo

59.8 HENDERSON ISLAND FRUIT-DOVE *Ptilinopus insularis* [Ptilope de Henderson] L 23cm. No similar fruit-dove on Henderson I. ❋ Forest. ☉ Pi 1. R E.Pi

59.9 MANY-COLOURED FRUIT-DOVE *Ptilinopus perousii* [Ptilope de La Pérouse] L 23cm. ♂ is unmistakable. Nom. ♀ (**a**, most of the species range except Fiji and Tonga) from ssp *mariae* ♀ (**b**, Fiji and Tonga) by yellow, not purple vent. Cf. 59.1. ❋ Forest, plantations, parks. ♫ High *wooh wooh - - woohpupooh*—. ☉ Ton 9, A.Sa, Sa, Fi.

60 FRUIT-DOVES & DOVES

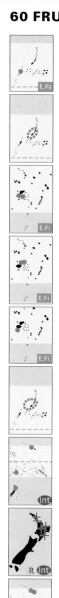

60.1 PALAU FRUIT-DOVE *Ptilinopus pelewensis* [Ptilope des Palau] L 24cm. No similar other bird species on Palau islands. ✳ Forest. ☉ Pa. E.Pa

60.2 MARIANA FRUIT-DOVE *Ptilinopus roseicapilla* [Ptilope des Mariannes] L 23cm. No similar other bird species on Mariana Is and Guam. ✳ Forest, second growth. ☉ Gu, NMa 1–4. R

60.3 GOLDEN DOVE *Ptilinopus luteovirens* [Ptilope jaune] L 20cm. ♂ unmistakable; ♀ from ♀ 59.9 by all-green plumage. Not on same island as 60.4. ✳ Forest, second growth. 60–2,000m. ☉ Fi 1,9–11,14. E.Fi

60.4 ORANGE DOVE *Ptilinopus victor* [Ptilope orange] L 19cm. ♂ unmistakable; ♀ from ♀ 59.9 by all-green plumage. ✳ Woodland, open forest. At altitudes between 420 and 980m. ☉ Fi 2,4,8,16,18,22. E.Fi

60.5 VELVET DOVE *Ptilinopus layardi* [Ptilope de Layard] L 20cm. ♂ unmistakable; ♀ from ♀ 59.9 by all-green plumage. ✳ Forest, lowland bush, village gardens. ☉ Fi 3,15. E.Fi

60.6 ISLAND COLLARED-DOVE *Streptopelia bitorquata* [Tourterelle à double collier] L 32cm. No similar dove in its range, except 60.7**b** with different black neck pattern and dark shafts of upperpart feathers. Note grey tail tips and orange upper mantle. ✳ Mainly mangroves, but also in other types of woodland and even in gardens. ☉ Gu, NMa 1–4.

60.7 SPOTTED DOVE *Streptopelia chinensis* [Tourterelle tigrine] L 29cm. Shown are Nom. (**a**, with uniform mantle and wing coverts, I to Hawaii islands) and ssp *tigrina* (**b**, with black shaft streaks to feathers of upperparts, I to Fiji and NI of NZ). ✳ Urban and suburban areas. ♫ High *pupu-proooh* (*prupru* hurried and *proooh* slightly descending). ☉ Fi, NZ 1, Ha 1–7. I R

60.8 AFRICAN COLLARED-DOVE[68] *Streptopelia roseogrisea* [Tourterelle rieuse] L 26cm. From 60.7 by paler plumage and different by patterned neck collar. ✳ Suburban areas, orchards. ♫ Mellow *p'prrooh*. ☉ NZ 1. I

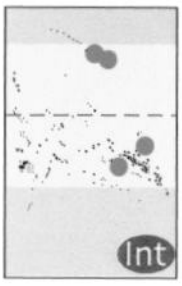

60.9 ZEBRA DOVE *Geopelia striata* [Géopélie zébrée] L 21cm. Unmistakable by barring and rufous wing panels. ✳ Dry woodland, cultivation, gardens. ♫ High, hoarse *prrrruh-ppppruh* and *puuhperrup*. ☉ FrPo 4,20, Ha (in Ha shares range with 60.10). I

60.10 MOURNING DOVE *Zenaida macroura* [Tourterelle triste] L 31cm. Note black spots on scapulars and tertials, attenuated tail, black spot below cheek and lack of black neck markings. ✳ Ranchland. ♫ Plaintive, low *oooh oooh oooh uwóoh*. ☉ Ha 2,3,5–7. I (no map)

61 PARAKEETS, PARROTS & ROSELLAS

61.1 BURROWING PARAKEET *Cyanoliseus patagonus* [Conure de Patagonie] L 45cm. Macaw-like. Note bright yellow rump. ✻ Nests and rests in sea cliffs. ☉ Ha 1. I

61.2 CRIMSON SHINING-PARROT *Prosopeia splendens* [Perruche pompadour] L 45cm. Head and underparts red, not maroon. ✻ Forest, mangrove, second growth, cultivation. ☉ Fi 1,3,15. E.Fi

61.3 RED SHINING-PARROT[69] *Prosopeia tabuensis* [Perruche écarlate] L 45cm. Maroon head and underparts diagnostic. Shown are ssp *taviunensis* (**a**, always without blue collar, Taveuni, Quamea) and other ssps/variations (**b**, with blue collar of varying width and with/without maroon spots to rump, Vanua Levu, Kioa and Koro). ✻ Mangrove, forest. ♫ Typical parrot-like *wràh wràh*. ☉ Fi 2,4,10,16–18, Ton 2(I). E.Fi

61.4 MASKED SHINING-PARROT *Prosopeia personata* [Perruche masquée] L 47cm. No similar bird in range. ✻ Forest, second growth, cultivation, mangroves. ☉ Fi 1. E.Fi

61.5 ECLECTUS PARROT *Eclectus roratus* [Grand Éclectus] L 39cm. Unmistakable, especially red-and-blue ♀. ✻ Most types of forested and wooded habitats. ☉ Pa 2–4. I

61.6 ROSE-RINGED PARAKEET *Psittacula krameri* [Perruche à collier] L 40cm. Note black-and-pink collar, red bill and black flight feathers if seen from below. ✻ Dry scrub, cultivation, orchards. ♫ Squeaky *sreew*. ☉ Ha 1,2,6,7. I

61.7 RED-CROWNED PARROT *Amazona viridigenalis* [Amazone à joues vertes] L 33cm. Note pale bill, red cap and black tips to neck and mantle feathers. ✻ Forest, orchards, parks. ☉ Ha 6. I

61.8 CRIMSON ROSELLA *Platycercus elegans* [Perruche de Pennant] L 37cm. Ad. and Imm. unmistakable. Imm. (shown in flight) from Imm. 61.9 (not shown) by white not blue throat. ✻ Suburban areas. ♫ Low, rapid, nasal, slightly descending *whetwhetwhet* and budgerigar-like twitters. ☉ NZ ?1.

61.9 EASTERN ROSELLA *Platycercus eximius* [Perruche omnicolore] L 31cm. Unmistakable by colour pattern of plumage. ✻ Woodland, woodland remains, orchards, cultivation. ♫ Nasal twittering, chatters and other calls and songs similar to 61.8. ☉ NZ 1,2. I

62.1 COLLARED LORY *Phigys solitarius* [Lori des Fidji] L 20cm. No similar species in range. ❋ Mainly in forest, but also in cultivation and suburban areas. ☉ Fi thr., except S islands of 7. E.Fi

62.2 POHNPEI LORIKEET *Trichoglossus rubiginosus* [Loriquet de Ponapé] L 24cm. No similar species in range. ❋ Forest, plantations. ☉ Mi 2. E.Mi

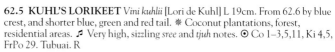

62.3 RED-THROATED LORIKEET *Charmosyna amabilis* [Lori à gorge rouge] L 19cm. Not together on same islands in Fiji as 62.4. Red thighs often concealed between feathers of lower belly. ❋ Forest canopy. ☉ Fi 1,2,4,11. R E.Fi

62.4 BLUE-CROWNED LORIKEET *Vini australis* [Lori fringillaire] L 19cm. From 62.3 by blue crown and different range. ❋ Plantations, gardens, scrub, forest. ☉ Ton, Ni, WaF, Fi 7 (S islands), Sa, A.Sa 4.

62.5 KUHL'S LORIKEET *Vini kuhlii* [Lori de Kuhl] L 19cm. From 62.6 by blue crest, and shorter blue, green and red tail. ❋ Coconut plantations, forest, residential areas. ♪ Very high, sizzling *sree* and *tjuh* notes. ☉ Co 1–3,5,11, Ki 4,5, FrPo 29. Tubuai. R

62.6 STEPHEN'S LORIKEET *Vini stepheni* [Lori de Stephen] L 19cm. Cf. 62.5. Only found on Henderson I. ❋ Forest. ♪ Very high *sreeeeh*. ☉ Pi 1. R E.Pi

62.7 BLUE LORIKEET *Vini peruviana* [Lori nonnette] L 18cm. Unmistakable by dark blue, almost black plumage with white bib. ❋ Any type of wooded habitat, but especially in coconut trees; also in low, flowering plants. ♪ Varied e.g. high, rapid *siesiesie*. ☉ FrPo 2,21,22, Co 5.

62.8 ULTRAMARINE LORIKEET *Vini ultramarina* [Lori ultramarin] L 18cm. No similar species in range. ❋ Any wooded habitat with flowering trees. ♪ Very high, thin, sizzling *srisisi*. ☉ FrPo 12,13,23. R E.FrPo

63 PARAKEETS, COCKATOO, GALAH, KEA, KAKA & KAKAPO

63.1 MITRED PARAKEET *Aratinga mitrata* [Conure mitrée] L 35cm. Note irregular bordered red cheeks, green underwing coverts and scattered red feathers. Normally with sparse or without red at wing bend. ✽ Residential areas with tall trees, sea cliffs. ☉ Ha 1,2. I

63.2 RED-MASKED PARAKEET *Aratinga erythrogenys* [Conure à tête rouge] L 33cm. Note red at wing bend especially when seen from below. Oahu flock may include a few 63.3. ✽ Forest, residential areas. ☉ Ha 6. I

63.3 BLUE-CROWNED PARAKEET *Aratinga acuticaudata* [Conure à tête bleue] L 36cm. Note blue at head, red upper mandible and black lower mandible. ✽ Forest, residential areas. ☉ ?Ha 6. I

63.4 SULPHUR-CRESTED COCKATOO *Cacatua galerita* [Cacatoès à huppe jaune] L 50cm. Unmistakable. ✽ Forest, plantations, farmland, orchards. Feeds mainly on the ground. ♫ Harsh, raucous and guttural screeches. ☉ NZ 1,2, Pa 1–4. I

63.5 GALAH *Eolophus roseicapilla* [Cacatoès rosalbin] L 36cm. Unmistakable. ✽ Open woodland, savanna, shrub, agricultural land. ☉ NZ 1. I

63.6 KEA *Nestor notabilis* [Nestor kéa] L 48cm. Only on SI, rare and almost exclusively restricted to mountains. From darker and smaller 63.7 by darker cap, and lack of crimson feather edges to feathers of vent and lower belly. ✽ Wide variety of habitats, especially at higher altitudes up to 2,100m, from meadows to temperate rainforest, rarely even seen in coastal flats. ♫ High, slightly descending, slightly mewing *keeaaah* —. ☉ NZ 1,2. R E.NZ

63.7 NEW ZEALAND KAKA[70] *Nestor meridionalis* [Nestor superbe] L 45cm. 2 ssps, *septentionalis* (**a**, with pale brown cap) and white-capped Nom. (**b**). Note yellow-orange cheek feathers and orange wash to neck feathers. NZ thr. but rare, except on a few islands. ✽ Prefers undisturbed native forest. Up to 1,200m. ♫ Varied e.g. scratchy *kraakaha*, melodious, whistled *peerowuh'peeruwuh* or liquid *lokloklok-*. ☉ NZ 1–3. R E.NZ

63.8 KAKAPO *Strigops habroptila* [Strigops kakapo] L 64cm. Strictly nocturnal and very rare. Flightless. Unmistakable in the unlikely event of being seen. ✽ Originally in a wide range of habitats: mossy forest, scrub, tussock grassland, peatlands, pastures. ♫ Very low, resounding, mechanical *booh booh* - -. ☉ NZ ?1,?2,?3 (probably extinct on mainland, re-introduced on several islands). E.NZ

64.1 ANTIPODES ([NZ]Island) **PARAKEET** *Cyanoramphus unicolor* [Perruche des Antipodes] L 30cm. Unmistakable in range by wholly green head. ❀ Tussock grassland, fernland, coastal swamp. ♫ Varied, rapid chattering, hoarse screeches, cooing squeaks. ☉ NZ 8. R E.NZ

64.2 YELLOW-FRONTED (or [NZ]-crowned) **PARAKEET** *Cyanoramphus auriceps* [Perruche à tête d'or] L 23cm. Note yellow forecrown; from 64.4 by range; from 64.5 by slightly broader, deeper orange forehead. ❀ Mainly in undisturbed native forest. ♫ Very high, shivering chatter and other calls (higher pitched than those of 64.3). ☉ NZ 1–3,?5. E.NZ

64.3 RED-FRONTED (or [NZ]-crowned) **PARAKEET**[71] *Cyanoramphus novaezelandiae* [Perruche de Sparrman] L 27cm. Shown are Nom. (**a**), *chathamensis* (**b**), *hochstteri* (**c**, differing from **a** and **b** by saturation of colours) and *cyanurus* (**d**, from **a** by more blue on wings and more bluish-green tail). ❀ Rainforest, scrubland, open areas. ♫ Varied e.g. very high, hoarse chattering *ki-ki-ki*— and sharp screeches. ☉ NZ 1–6,10. E.NZ

64.4 CHATHAM ISLANDS (or [NZ]Forbes') **PARAKEET** *Cyanoramphus forbesi* [Perruche des îles Chatham] L 26cm. Restricted range. From 64.3**d** by yellow forehead. ❀ Undisturbed forest, scrub. ☉ NZ 4. E.NZ

64.5 MALHERBE'S (or [NZ]Orange-fronted) **PARAKEET** *Cyanoramphus malherbi* [Perruche de Malherbe] L 20cm. Note restricted range. Cf. 64.2. ❀ Forest edge. ♫ Sharp chatter and a large diversity of musical and harsh notes, uttered singly or in series. ☉ NZ 2. E.NZ

64.6 DOLLARBIRD *Eurystomus orientalis* [Rolle oriental] L 30cm. Unmistakable. ❀ Prefers open forest and woodland. ♫ Unstructured series of chattered, raucous notes. ☉ NZ 1,2, Mi 2,4, Pa. R

64.7 RAINBOW BEE-EATER *Merops ornatus* [Guêpier arc-en-ciel] L 20cm, excl. streamers of up to 7cm. Unmistakable. ❀ Open forest, open woodland, scrub and residential areas. ☉ Pa, NMa 1.

64.8 EURASIAN HOOPOE *Upupa epops* [Huppe fasciée] L 27cm. Unmistakable. ❀ Forages in short grass in bush and woodland. ☉ NMa 1. V

65.1 CHESTNUT-WINGED CUCKOO *Clamator coromandus* [Coucou à collier] L 46cm. Unmistakable. ❃ Woodland, mangrove, scrub, cultivation. ☉ Pa. V

65.2 COMMON CUCKOO *Cuculus canorus* [Coucou gris] L 33cm. From 65.4 by denser barring below and whiter vent. Most ♀ ♀ grey like ♂, rufous morph (**a**) uncommon. ❃ Open forest, clearings, reedbeds, wooded grassland. ♫ Silent outside breeding season. ☉ Pa, Ha 17. R

65.3 MALAYSIAN HAWK-CUCKOO *Cuculus fugax* [Coucou fugitif] L 29cm. Vagrant to Palau. Not clear which ssp involved; probably *hyperythrerus* (**b**), but Nom. (**a**, strongly ressembling Imm. **b**) also possible. Note rufous subterminal tail bar, white neck spot and white feathers at lower back. ❃ Forest, bamboo thickets, plantations. ☉ Pa. V

65.4 ORIENTAL CUCKOO *Cuculus optatus* [Coucou asiatique] L 31cm. ♂ vent often tawny and unbarred. The dark bars on back of the rufous morph of ♀ (**a**) are normally broader than intermediate rufous bars (65.2 shows the reverse). ❃ Forest, woodland, mangrove, wooded farmland, gardens. ♫ Rapid, mid-high, hollow *poohpooh poohpooh poohpooh* (3–5 times). ☉ NMa, Mi 2,4, Pa, NZ 1,2,5. R

65.5 PALLID CUCKOO *Cuculus pallidus* [Coucou pâle] L 32cm Unmistakable. Note absence of barring to breast and belly. ❃ Open woodland, wooded farmland. ♫ Very high, fluted, rapidly ascending series of about 10 *puh* notes, the 1st stuttered. ☉ NZ 1,2. R

65.6 BRUSH CUCKOO *Cacomantis variolosus* [Coucou des buissons] L 25cm. Strongly ressembling 65.7, but not in same range. ❃ Forest, mangrove, plantations, cultivation. ☉ Pa. V

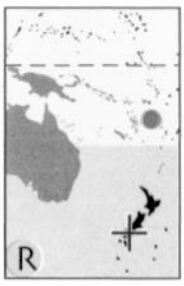

65.7 FAN-TAILED CUCKOO *Cacomantis flabelliformis* [Coucou à éventail] L 26cm. Unmistakable in range. Shown are ssp *pyrropanus* (**a**, V to NZ) and ssp *simus* (**b**, normal morph like **a**; less common black morph shown). ❃ Lower levels of forest and woodland. ♫ Short, high, whistled, descending trill. ☉ Fi 1,3,4,11, V NZ.

65.8 YELLOW-BILLED CUCKOO *Coccyzus americanus* [Coucou à bec jaune] L 30cm. Unmistakable. Note bicoloured bill and rufous in wings. ❃ From forest to arid scrub. ♫ Very high, nasal *tjark-tjark- -* (8 times) or bouncing *tjitjitji—*. ☉ Ha 13. V

65.9 SHINING BRONZE-CUCKOO[72] *Chrysococcyx lucidus* [Coucou éclatant] L 17cm. Unmistakable. Note small size; tail not attenuated. ❃ Forest, plantations, wooded farmland, suburban areas. ♫ Long, very high series of plaintive notes, starting with *tjuw-tjuw-tjuw*. ☉ NZ 1–6,10, Mi (Kapingamarangi I, not indicated on map). NZ

66 CUCKOO, KOELS, NIGHTJAR, NIGHTHAWK, KOOKABURRA & KINGFISHERS

66.1 CHANNEL-BILLED CUCKOO *Scythrops novaehollandiae* [Coucou présageur] L 60cm. Unmistakable by large size, long tail and huge bill. ✹ May be found in forest, woodland, wooded farmland. ♪ Mid-high *rararara* and other, short, raucous series, sometimes uttered in duet. ☉ NZ 1,3. V

66.2 AUSTRALIAN (or ᴺᶻPacific) **KOEL** *Eudynamys cyanocephalus* [Coucou bleuté] L 43cm. Vagrant to SI. ♀ ♀ and Imms could be confused with 66.3, but 66.2 is barred, not striped below and shows different head pattern. ♂ unmistakable (note size and red eyes). Dark ♀ morph (**a**) and pale ♀ morph (**b**) known. ✹ Wooded habitats. ☉ NZ 2. V

66.3 LONG-TAILED KOEL (or ᴺᶻCuckoo) *Eudynamys taitensis* [Coucou de Nouvelle-Zélande] L 40cm. Note barred and spotted upperparts and streaked underparts (unlike any other cuckoo). ✹ Forest canopy, plantations, suburban areas. ♪ Very high, hoarse, slightly upslurred *sreeeeuw'ft*. ☉ Thr., but not in Ha.

66.4 GREY NIGHTJAR *Caprimulgus indicus* [Engoulevent jotaka] L 30cm. Unmistakable in range. Longer-tailed and more nocturnal than 66.5. ✹ Forest, mangrove, cultivation, plantations. ♪ Regular, mechanical *chuk-chuk-chuk—*. ☉ Pa. *Note*: recently split into 2 species: INDIAN NIGHTJAR C. *indicus* (migrant from Asia) and GREY NIGHTJAR C. *jatoka* (E to Palau). Resemble each other closely.

66.5 COMMON NIGHTHAWK *Chordeiles minor* [Engoulevent d'Amérique] L 24cm. Unmistakable; may be seen by day. ✹ Open areas. ♪ Very high, hoarse *sreeuw sreeuw - -*. ☉ Ha 11. V

66.6 LAUGHING KOOKABURRA *Dacelo novaeguineae* [Martin-chasseur géant] L 41cm. Unmistakable. ✹ Wooded farmland, settlement. ♪ High, sharp, rapid chattering uttered in chorus. ☉ NZ 1,2. I

66.7 BELTED KINGFISHER *Megaceryle alcyon* [Martin-pêcheur d'Amérique] L 31cm. Unmistakable. ✹ Coastal bays, estuaries. ☉ Ha 1–7.

66.8 COMMON KINGFISHER *Alcedo atthis* [Martin-pêcheur d'Europe] L 16cm. Rufous cheeks diagnostic. ✹ At clear, slowly flowing waters with some overhanging vegetation. ♪ Extreme high *feet feet-feet*. ☉ Gu. V

67.1 CHATTERING KINGFISHER *Todiramphus tutus* [Martin-chasseur respecté] L 22cm. 3 ssps known: Nom. (**a**, in FrPo sole kingfisher on many islands except on Tahiti, where together with very different 67.7), ssp *atiu* (**b**, sole kingfisher on Atiu I.) and ssp *mauke* (not shown, blue on crown intermediate between **a** and **b**; sole kingfisher on Mauke I.). ✳ Highland streams in forest, second growth, plantations, gardens. ☉ FrPo 4,14–18,24, Co 3,11, Fi 7, Ton 1,2, A.Sa 3,4.

67.2 MANGAIA KINGFISHER *Todiramphus ruficollaris* [Martin-chasseur de Mangaia] L 22cm. Sole kingfisher on Mangaia I. ✳ Woodland, scrub. ♫ Loud, high *tjewtjewtjew*—. ☉ Co 2. R E.Co

67.3 SACRED KINGFISHER[73] *Todiramphus sanctus* [Martin-chasseur sacré] L 22cm. From 67.4 by range. ✳ Forest, tall woodland, mangrove. ♫ Loud, sharp, *wekwekwek-*. ☉ NZ 1–4,10.

67.4 COLLARED KINGFISHER *Todiramphus chloris* [Martin-chasseur à collier blanc] L 24cm. Many ssps exist from Africa and Asia to Pacific, of which 11 in area. In a few cases ♂ and ♀ differ as shown for ssp *vitiensis* (**e**, Fiji). Also as examples shown ssps *owstoni* (**a**, N NMa), *pealei* (**b**, Tutuila I.), *regina* (**c**, Futuna) and *albicilla* (**d**, S NMa). ✳ Wide variety of habitats, mangrove, gardens, road sides, savanna. ☉ ?Pa, ?Mi 4, Ma, Fi, WaF 1.

67.5 FLAT-BILLED KINGFISHER *Todiramphus recurvirostris* [Martin-chasseur des Samoa] L 22cm. Sole kingfisher on W Samoa Is. ✳ Wide variety of habitats, from montane forest to suburban areas. ☉ Sa. E.Sa

67.6 MICRONESIAN KINGFISHER *Todiramphus cinnamominus* [Martin-chasseur cannelle] L 20cm. Shown are ssps *pelewensis* (**a**, Palau Is) and *reichenbachii* (**b**, Pohnpei). From 67.4 by absence or less extensive blue in rufous crown. ✳ Forest. ☉ Mi 2, Pa.

67.7 TAHITI KINGFISHER *Todiramphus veneratus* [Martin-chasseur vénéré] L 21cm. Shown are Nom. (**a**, dull-coloured ♂ with weak frontal collar; ♀ dark-brown) and ssp *youngi* (**b**, pale brownish, sole kingfisher on Moorea). ✳ Forest, plantations, second growth. ♫ Very high, rapid, sharp, trilling *teereereeh* in series of 3. ☉ FrPo 4,19. E.FrPo

67.8 TUAMOTU (or Niau) **KINGFISHER** *Todiramphus gambieri* [Martin-chasseur des Gambier] L 20cm. Variants with or without stripe behind eyes shown. From 67.9 by range. ✳ Woodland, plantations, villages. ☉ FrPo 30. R E.FrPo

67.9 MARQUESAS KINGFISHER *Todiramphus godeffroyi* [Martin-chasseur des Marquises] L 21cm. Sole kingfisher on S Marquesas. ✳ Dense forest, plantations. ☉ FrPo 25,?20,?26. E.FrPo

68 SWIFTLETS, NEEDLETAIL, SWIFT & TREESWIFT

68.1 WHITE-RUMPED SWIFTLET *Aerodramus spodiopygius* [Salangane à croupion blanc] L 11cm. 68.1–7 are very similar and are best separated by range. 68.1 has the palest rump. ✳ Over wide range of habitats incl. grasslands, roads and forest. ♫ High *weetweet* or *tjurreweet*. ⊙ A.Sa, Sa, Ton, WaF, Fi (excl. 5).

68.2 POLYNESIAN SWIFTLET *Aerodramus leucophaeus* [Salangane de la Société] L 11cm. Cf. 68.1. ✳ Over rivers and forested valleys. ⊙ FrPo 4,19. R E.FrPo

68.3 MARQUESAN SWIFTLET *Aerodramus ocistus* [Salangane des Marquises] L 11cm. Cf. 68.1. ✳ Forages flying through forest canopy and edges. ♫ E.g. high *tree-tuh*. ⊙ FrPo 12,13,23,25,26,28. E.FrPo

68.4 ATIU SWIFTLET *Aerodramus sawtelli* [Salangane de Cook] L 10cm. Cf. 68.1. Somewhat contrasting paler underparts. ✳ Over forest and open areas. ⊙ Co 3. R E.Co

68.5 CAROLINE ISLANDS SWIFTLET *Aerodramus inquietus* [Salangane des Carolines] L 11cm. Cf. 68.1. Note dark rump, slightly paler than mantle. ✳ Open and forested areas. ⊙ Mi 1,2,3,?4. E.Mi

68.6 MARIANA SWIFTLET *Aerodramus bartschi* [Salangane de Guam] L 11cm. Cf. 68.1. Darkest rump. ✳ Over forest, forest remains, mangrove. ⊙ Ha 6, NMa 1,3, Gu. R

68.7 PALAU SWIFTLET *Aerodramus pelewensis* [Salangane des Palau] L 11cm. Cf. 68.1. ✳ Over canyons. ⊙ Pa. R E.Pa

68.8 WHITE-THROATED NEEDLETAIL *Hirundapus caudacutus* [Martinet épineux] L 20cm. Unmistakable by dark-and-white plumage pattern and shape of wings. Tail needles only visible in hand. View from above (**a**) for 68.8–9 shown to smaller scale. ✳ Over any type of wooded habitat, sometimes over mudflats, around coastal cliffs or well out at sea. ♫ Shrill twittering. ⊙ NZ 1,2,4,5,7, Pa 5, Gu, Fi 19,20. R

68.9 FORK-TAILED SWIFT *Apus pacificus* [Martinet de Sibérie] L 18cm. Note white rump and scaled underparts. ✳ Mostly over open habitats, near cliffs and coastal habitats. ♫ Very high, drawn-out, descending *sreeeeeeuw*. ⊙ NZ 1,2,4, Ha 17, Ma 1, NMa, Gu. R

[68.10 MOUSTACHED TREESWIFT *Hemiprocne mystacea* [Hémiprocné à moustaches] L 30cm. Unmistakable. ✳ Forest edge and open country with some trees. ⊙ Tentative sighting on Fiji. ?V (no map).]

69.1 BARN OWL *Tyto alba* [Effraie des clochers] L 34cm. Very pale expecially in flight. Nocturnal and crepuscular. Normally does not roost on ground. Nests in tree holes, buildings, barns. ❋ Open areas from farmland and grassy woodland to suburbs. ♫ Dry, hoarse, menacing shriek. ☉ NZ 1,2, Ha 1–7, Fi, WaF, Ton, Sa, A.Sa, Ni. R[74]

69.2 AUSTRALASIAN (or Eastern) **GRASS-OWL** *Tyto longimembris* [Effraie de prairie] L 34cm. Smaller-eyed, longer-legged and normally darker-backed than 69.1. Normally nests and roosts on ground. Very variable from uncommon pale like 69.1 to much darker as shown; however, in the area only known as dark morph from Viti Levu, but not seen there since about 1861. ❋ Tussock grassland, marsh, sugarcane fields, paddy fields. ☉ ?Fi.

69.3 SHORT-EARED OWL *Asio flammeus* [Hibou des marais] L 37cm. Note yellow, dark-set eyes, contrast between streaked breast and thinly striped belly, small ear tufts. Also active during the day. Ssp *sandwichensis* (**a**, from Hawaii) darker. ❋ Open habitats. ♫ High barks as if from a small dog *wreh-wreh*. ☉ Gu, NMa 1,2,6, Mi 1,?2,4, Ma, Ha.

69.4 GREAT HORNED OWL *Bubo virginianus* [Grand-duc d'Amérique] L 51cm (♂), 60cm (♀). Unmistakable by size, thick-set jizz and ear tufts. ❋ Prefers open, wooded habitats. ☉ FrPo 20. I

69.5 MOREPORK[75] *Ninox novaeseelandiae* [Ninoxe boubouk] L 29cm. Darker, larger-headed and longer-tailed than 69.7. ❋ Most wooded habitats from rainforest to dry woodland. ♫ High bassoon-like *more-pork* with slight tremolo. ☉ NZ 1–3,5.

69.6 BROWN HAWK-OWL *Ninox scutulata* [Ninoxe hirsute] L 30cm. Large, wide-set eyes and uniform dark brown, unmarked upperparts diagnostic. ❋ Habitats with trees such as forest, mangrove, parks and suburbs. ☉ Pa 12, NMa 4.

69.7 LITTLE OWL *Athene noctua* [Chevêche d'Athéna] L 22cm. Cf. 69.5. May be seen by day. ❋ Open farmland, parks, gardens. ♫ Loud, strong, strident *weeuw*. ☉ NZ 2. I

69.8 PALAU OWL *Pyrroglaux podarginus* [Petit-duc des Palau] L 22cm. Sole owl on Palau. ❋ All types of forest. ☉ Pa. E.Pa

70.1 YELLOW WAGTAIL *Motacilla flava* [Bergeronnette printanière] L 16cm. Visiting ssps might include *taivana* (**a**) and *thunbergi* (**b**). Cf. 70.2. ❋ Pastures, damp grassland, margins of water bodies. ♫ Very high *zweep*. ☉ Gu, Mi 3,4, Pa, NMa 4.

70.2 EASTERN YELLOW WAGTAIL *Motacilla tschutschensis* [Bergeronnette de Béringie] L 16cm. From 70.1 by white eyebrow. *Note*: shown is Nom., but also ssp *simillima* might be involved, differing by heavier bill. ❋ Like 70.1. ☉ Gu, Mi 3,4, Pa, NMa 4. (70.1 and 70.2 share winter range.)

70.3 WHITE WAGTAIL *Motacilla alba* [Bergeronnette grise] L 17cm. Rare visitors might concern grey-backed ssp *ocularis* (**a**) or/and black-backed ssp *leucopsis* (**b**) or *lugens* (**c**). Shown are Br ♂♂ (**a**, **b** and **c**) with all-white wing coverts, which are only edged white in N-br ♂♂, ♀♀ and Imms. ❋ Wide variety of open habitats. ♫ Very high *tjírruk* or extreme high *tsissik*. ☉ ?Gu, Pa.

70.4 GREY WAGTAIL *Motacilla cinerea* [Bergeronnette grise] L 19cm. From 70.1–2 by yellow rump, more pronounced white wingbars, grey mantle, longer tail and by calls. ❋ Edges of fast flowing water, forest tracks, sewage ponds. ♫ Very high *twitwit weet*. ☉ Gu, Pa.

70.5 RED-THROATED PIPIT *Anthus cervinus* [Pipit à gorge rousse] L 15cm. Note heavily streaked upperparts. Ads show a varying area of pink colouring at head and breast. ❋ Short grassland, marshes, muddy edges of wetlands. ♫ High, thin *sièèh*. ☉ Pa, Mi 4, Ha 18.

70.6 OLIVE-BACKED PIPIT *Anthus hodgsoni* [Pipit à dos olive] L 16cm. Note dark mark on rear ear coverts, white eyebrows and poorly striped back. ❋ Wooded habitats. ☉ Ha 18. V

70.7 AMERICAN PIPIT *Anthus rubescens* [Pipit d'Amérique] L 16cm. Unknown which ssp is involved; shown is Br *japonicus* with pink legs. Nom. and ssp *alticola* have dark legs. N-br plumage of all ssps is normally slightly to heavily streaked brown and less pinkish-buff. ❋ Open habitats along water bodies, marshes, beaches, mudflats, bare fields. ♫ Extremely high *jit jirrit jit*. ☉ Ha 18.

70.8 AUSTRALASIAN (or ᴺᶻNew Zealand) **PIPIT**[76] *Anthus novaeseelandiae* [Pipit austral] L 18cm. No other pipit in New Zealand. Shown are: ᴺᶻNom. (**a**, NI and SI, hardly marked below), *chathamensis* (**b**, ᴺᶻChatham Island, almost white below) and *aucklandicus* (**c**, Auckland Is, more buff than **a**); ssp *steindachneri* (**d**, from Antipodes Is) resembles smaller Nom. ❋ Open habitats, occasionally in open woodland. ♫ *wswee* or descending *sreeeuw* or very high, sharp series *weet-witwitwitwit*. ☉ NZ 1–8,10.

70.9 WATER PIPIT *Anthus spinoletta* [Pipit spioncelle] L 16cm. Note (nearly) unstreaked upperparts and dark legs. ❋ Open coastal habitats, edges of water bodies, grassland. ☉ Ha 18. V

70.10 SKYLARK *Alauda arvensis* [Alouette des champs] L 18cm. Text and map opposite plate 71.

71 RIFLEMAN, NZ WREN, DUNNOCK, LAUGHINGTHRUSHES, HWAMEI & LEIOTHRIX

70.10 SKYLARK *Alauda arvensis* [Alouette des champs] L 18cm. Compact build with white tail sides and trailing edge of wings. Erectile crest (**a**, crest normally held flat). ✳ Natural and cultivated, grassy habitats such as tussock grassland, pastures, golf courses, heathland. ♫ Flight call: *pree-it*. Sustained, aerial, cheerful, warbling song, each note repeated and varied. ☉ Ha 1–8,17,18, NZ 1–6,8,10. I. Illustration on previous page.

71.1 RIFLEMAN[77] *Acanthisitta chloris* [Xénique grimpeur] L 8cm. Note white eyebrows, pale wingbars, white patch on tertials. NZ's smallest bird. ✳ Forest, pine plantations, locally in large hedgerows. Up to 1,550m. ♫ Very high-pitched, week *srit*, single or in series. ☉ NZ 1–3. E.NZ

71.2 SOUTH ISLAND (or [NZ]Rock) **WREN** *Xenicus gilviventris* [Xénique des rochers] L 9.5cm. Unmistakable in its range by very short tail and general colour pattern. Hops and runs, avoiding flying. ✳ Rocky habitats with low scrub, 920–2,900m. ♫ Extremely high (for some people inaudible) *seet-sit-sit*. ☉ NZ 2. E.NZ

71.3 DUNNOCK (or [NZ]Hedge Sparrow) *Prunella modularis* [Accenteur mouchet] L 14cm. At first glance resembles House Sparrow, but note slim, neat apparance, thin bill, grey plumage parts of head and white wing spots. ✳ Dense shrub, orchards, gardens, dense second growth. Up to 1,500m. ♫ Short, very high, cheerful, warbled *siwisiwi-tusisi-seewi*. ☉ NZ 1–8. I

71.4 GREATER NECKLACED LAUGHINGTHRUSH *Garrulax pectoralis* [Garrulaxe à plastron] L 30cm. Unmistakable by head and upper breast pattern. Note pale tips to outer tail feathers. ✳ Lowland forest. ♫ Flock call: rapid, descending *wutwutwutwut*. ☉ Ha 7.

71.5 GREY-SIDED LAUGHINGTHRUSH *Garrulax caerulatus* [Garrulaxe à flancs gris] L 28cm. Note black lores, white marks below eyes (varying in size and form) and fine scaling at crown and cheeks. ✳ Dense forest. ☉ Ha 6. I

71.6 HWAMEI *Garrulax canorus* [Garrulaxe hoamy] L 23cm. Unmistakable by pale blue eyering with short backward running streak and rather uniform warm brown plumage. Head and neck slightly streaked. ✳ Dense forest; also in parks and gardens. ♫ Call: high, shrill, sustained *sreeh-sreeh*—. Song: rapid, thrush-like fluting with repetitions in strophes of 5–6 sec. ☉ Ha 1–7. I

71.7 RED-BILLED LEIOTHRIX *Leiothrix lutea* [Léiothrix jaune] L 15cm. No similar bird on Hawaii islands. ✳ Thick undergrowth of forest, scrub, plantations, bamboo thickets. ♫ Alarm call: high, rapid, sharp *sritsritsrit*—. Song: e.g. variations of high *weet-weetohweet-weet*. ☉ Ha 1,2,5,6. I

72 TRILLERS, MINIVET, CICADABIRD, CUCKOO-SHRIKE, SHRIKE & MOCKINGBIRD

72.1 WHITE-WINGED TRILLER *Lalage tricolor* [Échenilleur tricolore] L 17cm. Vagrant from Australia. From 72.2 by range and inconspicuous eyebrow. Note the long upper- and undertail coverts of 72.1–6 ✳ Open forest, woodland, wooded farmland, shrubland. ♫ High, vigorous rattle, slightly descending at the end (10–15 sec). ⊙ NZ 2. R

72.2 POLYNESIAN TRILLER *Lalage maculosa* [Échenilleur de Polynésie] L 16cm. Ssps variable, according to island, where found; e.g. birds from Samoa, Tonga and Niue (**a**) are clean white-black (♂) or white-brown with some barring to breast sides and flanks (♀); birds from Fiji (**b**) have barred underparts and those from Rotuma (**c**) are buffy below. ✳ Forest, also in gardens and parks. ♫ High, loud, rapid *weetweetweet*— (5 sec) or *tutjew* and other sharp or melodious notes and strophes. ⊙ WaF, Ton, Ni, Sa, Fi.

72.3 SAMOAN TRILLER *Lalage sharpei* [Échenilleur des Samoa] L 13cm. Unmistakable by whitish eyes, orange bill and lack of white in wings. No similar bird in range. ✳ Forest, wooded farmland. ♫ Short, nasal, upslurred squeaks. ⊙ Sa. R E.Sa

72.4 ASHY MINIVET *Pericrocotus divaricatus* [Miniver cendré] L 18cm. Note slender and long-tailed jizz. Small white patch at opened wing, especially visible from below. ✳ Wide variety of habitats, incl. forest, woodland, plantations, gardens. ⊙ Gu.

72.5 CICADABIRD *Coracina tenuirostris* [Échenilleur cigale] L 26cm. ♂ from larger and paler 72.6 by range and less solid black mask. ♂♂ rather similar within range but ♀♀ differ per island, *nesiotis* (**a**, from Yap) is more rufous and larger than heavily barred *monacha* (**b**, from Palau), while *inspirata* (**c**) from Pohnpei is unbarred with slaty head and cinnamon-rufous body. ✳ Forest, woodland, scrub, savanna, mangrove. ⊙ Mi 2,4, Pa.

72.6 BLACK-FACED CUCKOO-SHRIKE *Coracina novaehollandiae* [Échenilleur à masque noir] L 34cm. Occasionally stragglers from Tasmania found in New Zealand; distinctive by all-grey plumage and black mask; black to chin missing in Imm. ✳ Forest edge and clearings, woodland, parks and gardens, mangrove, wooded farmland. ♫ E.g. very rapid *T'SW'Rueeh* or *Tu'd'wir*. ⊙ NZ 1,2. R

72.7 BROWN SHRIKE *Lanius cristatus* [Pie-grièche brune] L 19cm. Asian straggler to Palau. Both ssps, reddish-brown *superciliosus* (**a**, ♀ often with some scaling to body sides as shown) and olive-green *lucionensis* (**b**) might have wandered in from Japan or from elsewhere in NE Asia. Black mask in all plumages distinctive. ✳ Wide variety of habitats such as open country, clearings. Often on telephone lines. ⊙ Pa.

72.8 NORTHERN MOCKINGBIRD *Mimus polyglottos* [Moqueur polyglotte] L 26cm. Introduced on Ha Is. Note long tail with white outer feathers; in flight with conspicuous white wing flashes. ✳ Dry, brushy habitats. ♫ *tjiptjip*— (10 times), —*tuweetuwee*— (4 times), —*weh*-rattle etc., each part well-accentuated. ⊙ Ha. I

73 RUBYTHROAT, ROCK-THRUSH, THRUSHES & SHAMA

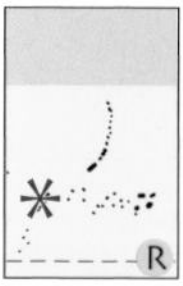

73.1 SIBERIAN RUBYTHROAT *Luscinia calliope* [Rossignol calliope]
L 15cm. Note in all plumages white (or pale) eyebrow, dark lores and short tail.
Skulking. ✳ Dense growth such as thickets, scrub, hedges, bamboo brakes, long
grass, reeds, gardens. ☉ Pa.

73.2 BLUE ROCK-THRUSH *Monticola solitarius* [Monticole merle-bleu]
L 22cm. ♂ distinctive by dull blue plumage with brown wings and tail; note that
Br ♂ of ssp *philippensi* (winter visitor to Palau) shows rufous underparts, but is
barred black-and-grey in N-br plumage. ♀ ♀ are densely scalloped below.
✳ Any habitat with some bare rock or buildings. ☉ Pa.

73.3 EYEBROWED THRUSH *Turdus obscurus* [Merle obscur] L 22cm.
Distinctive white eyebrow and lower eyelid. Note black lores. ♂ shows grey
head. Underwings greyish white. ✳ Winters in a wide variety of wooded habitats
from open forest, plantations and orchards to mangrove. ☉ Ha 17, Pa.

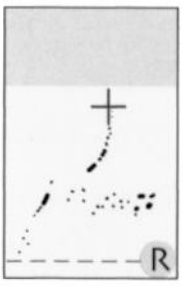

73.4 DUSKY THRUSH *Turdus naumanni eunomus* [Grive de Naumann]
L 24cm. Note rufous wings and scalloped plumage. Wings all-rufous below.
✳ Winters in open wooded areas such as scrub, wooded farmland, orchards,
suburbs. ☉ NMa 10,11.

73.5 EURASIAN BLACKBIRD *Turdus merula* [Merle noir] L 26cm. Rather
large thrush, ♂ unmistakable by black plumage and yellow bill; ♀ ♀ are brown
with paler chin. ✳ Habitats with scrub, such as gardens, farmland, shrubland,
heath, tussock grassland, mangrove. Up to 1,500m. ♫ Loud, very melodious,
continuous stream of well-separated strophes, all with the same structure but
never two the same. ☉ NZ 1–8,10. I

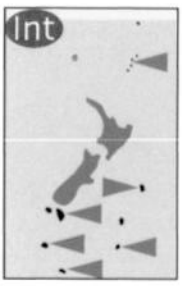

73.6 SONG THRUSH *Turdus philomelos* [Grive musicienne] L 22cm. No
similar thrush in range. ✳ Like 73.5. ♫ Incessant stream of loud, short,
melodious strophes, each one containing three–four repeated notes or group of
notes. ☉ NZ 1–8,10. I

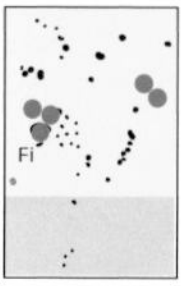

73.7 ISLAND THRUSH *Turdus poliocephalus* [Merle des îles] L 20–23cm. Very
variable depending on island where found. Yellow-orange bill, eyering and legs
diagnostic. Six ssps shown: *hades* (**a**, Fi 10), *ruficeps* (**b**, Fi 3), *vitiensis* (**c**, Fi 2),
tempesti (**d**, Fi 4), *layardi* (**e**, Fi 1,6,11,17) and *samoensis* (**f**, Sa). ✳ Dense forest.
☉ Fi 1–4,6,10,11,17, Sa.

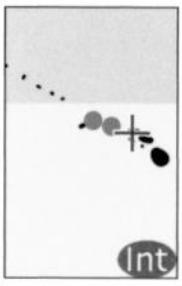

73.8 WHITE-RUMPED SHAMA *Copsychus malabaricus* [Shama à croupion
blanc] L 25cm (incl. tail). Unmistakable by long tail and white rump.
✳ Undergrowth of forest, thickets, gardens. ♫ Calm, thrush-like, slightly
descending cocktails of fluted, melodious, shrill and repeated notes in phrases of
5–10 sec. ☉ Ha 5–7.

74.1 OMAO *Myadestes obscurus* [Solitaire d'Hawaï] L 19cm. Note grey forehead. 74.1–4 are endemic thrushes, related to the N American Solitaire. They resemble each other very closely but their ranges do not overlap, except 74.2 and 74.3. Imms (74.1 given as example) are spotted above and scalloped below. ❀ Forest, scrub and savanna above 1,000m and open scrub above 2,000m on Mauna Loa. ♪ Hoarse *wrèèèh* or high police whistle-like *rréeh*. Song is liquid, rapid, varied *weetjohwih*. ☉ Ha 1. E.Ha

74.2 KAMAO *Myadestes myadestinus* [Solitaire kamao] L 19cm. See 74.3. From 74.1 mainly by brown forehead. ❀ Dense montane forest. ♪ Call: raspy *sreeew* or police whistle-like *peeeh*. Song: series of calm short trills and fluted notes, *piu-pi-uh-wreeh*. ☉ ?Ha 7. E.Ha

74.3 PUAIOHI *Myadestes palmeri* [Solitaire puaïohi] L 17cm. From larger 74.2 by different face pattern and pink legs. ❀ At streams in undergrowth of dense forest. ♪ Call: very high, sizzling *ssssh*. Song: high, short, descending phrases, *ohcomenowyou*. ☉ Ha 7. R E.Ha

74.4 OLOMAO *Myadestes lanaiensis* [Solitaire de Lanai] L 17cm. Buff patch at base of primaries more distinct than in other thrushes. ❀ Montane forest. ☉ Ha ?5. R E.Ha

74.5 RED-VENTED BULBUL *Pycnonotus cafer* [Bulbul à ventre rouge] L 20cm. Note dark hood, white rump and red undertail coverts. ❀ Forest, farmland, towns, villages. ♪ High *prueeh* or *preeh-preeh* or low liquid *witwit* or hoarse, yet melodious *sreetewtew*. ☉ FrPo 4, Ton, Fi, Ha 6, Ma 12, ?NZ, Sa, A.Sa. I

74.6 RED-WHISKERED BULBUL *Pycnonotus jocosus* [Bulbul orphée] L 19cm. Unmistakable by long upright crest and facial pattern. Imm. lacks red in cheeks. ❀ Suburban areas. ♪ Short, very high, hurried, cheerful phrases. ☉ Ha 6. I

74.7 LANCEOLATED WARBLER *Locustella lanceolata* [Locustelle lancéolée] L 12cm. Note short-tailed jizz, short dark streaks to breast and all-dark, pale-rimmed tertials. Creeps mouse-like low down through grass. ❀ Damp, grassy habitats such as marshes, scrub at reedbeds, vegetated ditches, rice paddies. ☉ Pa, NMa.

74.8 LONG-LEGGED WARBLER *Trichocichla rufa* [Mégalure des Fidji] L 19cm. Dark rufous with prominent eye-stripe. Terrestrial beneath dense vegetation. ❀ Rainforest above 800m. ☉ Fi 1,2. R E.Fi

74.9 FERNBIRD[78] *Megalurus punctatus* [Mégalure matata] L 18cm. No similar bird in range. Skulking in dense cover; weak flyer; supports itself on tail when clambering around. Six ssps (as examples Nom. **a** and ssp *wilsoni* **b** shown), differing in size and saturation of colours. ❀ Dense low vegetation in fresh and saline wetlands. ♪ Contact call is duet, started by ♂, answered by ♀, together as *TU'Weeh*. Also heard is rapid *Turrit* or *Trit*, short, dry rattle or fluted *Tuwt*. ☉ NZ 1–3. E.NZ

75 REED-WARBLERS, MILLERBIRD & BUSH WARBLERS

75.1 PALAU BUSH-WARBLER *Cettia annae* [Bouscarle des Palau] L 15cm. Note white eyebrow, pale orange legs and long, thin bill. ✤ Dense forest undergrowth. ☉ Pa. E.Pa

75.2 JAPANESE BUSH-WARBLER *Cettia diphone* [Bouscarle chanteuse] L 17cm. Very skulking. Note rufous tone to forehead, flight- and tail feathers. ✤ Dense forest undergrowth and bamboo thickets. Often in suburban gardens. ♫ Call: high, short *chip*. Song: very high, hurried phrases, introduced by two to three slow, melodious notes. ☉ Ha 1–7. I

75.3 FIJI BUSH-WARBLER *Cettia ruficapilla* [Bouscarle des Fidji] L 13cm. Several races, of these *badiceps* (**a**, without rufous in cheeks, Viti Levu), Nom. (**b**, Kadavu) and *funebris* (**c**, Taveuni) shown. All races have dark rufous crown and pale pink legs. Nom. with indistinct eyebrow. All races rather dark above with rufous tinge, especially to wing and tail feathers. ✤ Dense forest undergrowth and second growth. ☉ Fi 1–4. E.Fi

75.4 GREAT REED-WARBLER *Acrocephalus arundinaceus* [Rousserolle turdoïde] L 20cm. Note strong bill. May raise crest. ✤ Reeds, marsh, sugarcane, thickets, forest clearings. ♫ Mid-high, loud, hoarse *karre-keét-weét*. ☉ ?Pa (probably erroneous record).

75.5 ORIENTAL REED-WARBLER *Acrocephalus orientalis* [Rousserolle d'Orient] L 18cm. Rufous-toned, olive upperparts. Note faint striping to lower throat and upper breast. ✤ Reed-beds, tall grass, bush. ☉ Pa, ?Mi.

75.6 AUSTRALIAN REED-WARBLER *Acrocephalus australis* [Rousserolle d'Australie] L 17cm. Resembles larger 75.5, but with shorter bill. ✤ Low weedy vegetation in and around any type of wetland. ☉ NZ 2.

75.7 NIGHTINGALE REED-WARBLER *Acrocephalus luscinius* [Rousserolle rossignol] L 18cm. Unmistakable by extremely long and thin bill. ✤ Reed-beds, thickets, forest undergrowth. ☉ NMa 1,3,6,12, ?Gu, Mi ?1,2,4, Na. R

75.8 MILLERBIRD *Acrocephalus familiaris* [Rousserolle obscure] L 13cm. No similar bird in its range. ✤ Bushy hillsides. ♫ High, nasal, sustained *cheep cheep cheep* with many variations. ☉ Ha 8. E.Ha

75.9 CAROLINE REED-WARBLER *Acrocephalus syrinx* [Rousserolle des Carolines] L 15cm. Rather dark overall with long bill (but not as long as 75.7). ✤ Forest, second growth, gardens, stands of tall grass. ☉ Mi 2–4. E.Mi

76.1 TAHITI REED-WARBLER *Acrocephalus caffer* [Rousserolle à long bec] L 18cm. Unmistakable in range. Distinctively long bill. Note melanistic morph (**a**). ❊ Forest, second growth, bamboo thickets, coconut plantations. ♫ Series of high, single, strong *tjew* notes alternated with melodic *weetjewhat*. ⊙ FrPo 4,15,19. R E.FrPo

76.2 MARQUESAN REED-WARBLER *Acrocephalus mendanae* [Rousserolle des Marquises] L 18cm. Unmistakable in range. ❊ Forest, plantations, brush. Up to 1,200m. ♫ Strange, rapid series of alternating short, melodious notes, hoarse croaks and high-pitched twitters. ⊙ FrPo 3. E.FrPo

76.3 CHRISTMAS ISLAND WARBLER *Acrocephalus aequinoctialis* [Rousserolle de la Ligne] L 15cm. Unmistakable in range. Note pale grey plumage. Only unmated ♂♂ sing. ❊ Dense brush, open, sparsely wooded areas. ♫ Low, hoarse, rather toneless *srrruh*. ⊙ Ki 4,5,13. E.Ki

76.4 TUAMOTU REED-WARBLER *Acrocephalus atyphus* [Rousserolle des Touamotou] L 18cm. Several ssps, of these yellow-bellied *flavidus* (**a**, Napuku), grey Nom. (**b**, NW Tuamoto Arch.) and cinnamon-tinted *eremus* (**c**, Makatea) shown. ❊ Woodland, bush, plantations, gardens. ♫ *sree*, preceded and followed by some mellow or two staccato, high, finch-like notes. ⊙ FrPo 2. E.FrPo

76.5 COOK ISLANDS REED-WARBLER *Acrocephalus kerearako* [Rousserolle des Cook] L 16cm. Unmistakable in range. ❊ Woodland, reeds, gardens. ♫ Call: *wreet* or *wrut* or *feetjeeh*. Song: energetic stream of high, loud notes and trills, often introduced by *twit*. Also series of well-separated, mid-high *chret* and high *weet* notes. ⊙ Co 2,9. E.Co

76.6 PITCAIRN REED-WARBLER *Acrocephalus vaughani* [Rousserolle des Pitcairn] L 17cm. Unmistakable in range. Extend of white varying in range (less white on Pitcairn, more so on Henderson and Rimatara). Yellow Imm. similar to Imms of several other species. ❊ Prefers tall forest. ♫ High, unmelodic *sreeeh sreeeh*. ⊙ Pi 2. R E.Pi

76.7 RIMATARA REED-WARBLER *Acrocephalus rimatarae* [Rousserolle de Rimatara] L 17cm. Unmistakable in range. ❊ Forest, reed-beds. ♫ *cheeck* notes in irregular series. ⊙ FrPo 27. R E.FrPo

76.8 HENDERSON ISLAND REED-WARBLER *Acrocephalus taiti* [Rousserolle d'Henderson] L 17cm. Unmistakable in range. Head normally all-white with some blackish feathers in crown. ❊ Forest with thick undergrowth. ♫ Unmelodic, scratchy *sruuuh* or *sreeeh*. ⊙ Pi 2. R E.Pi

76.9 NAURU REED-WARBLER *Acrocephalus rehsei* [Rousserolle de Nauru] L 15cm. 'Average', brownish reed-warbler, unmistakable in small range. ❊ Forest, scrub, gardens. ⊙ Na. R E.Na

77 GERYGONES, YELLOWHEAD, WHITEHEAD, PIPIPI, FLYCATCHERS & SHRIKEBILLS

77.1 GREY GERYGONE (or [NZ]Warbler) *Gerygone igata* [Gérygone de Nouvelle-Zélande] L 10cm. Small, greyish green, lacking eyebrow; distinctive tail pattern not easy to see in the field. ❋ All types of native wooded habitats from forest to shrublands and mangroves. ♫ Cheerful, very high, warbling song, slightly descending, based on *turreweet* (in which *tur* low- and *weet* very high-pitched). ☉ NZ 1–3,5. E.NZ

77.2 CHATHAM ISLAND GERYGONE[80] (or [NZ]Warbler) *Gerygone albofrontata* [Gérygone des Chatham] L 12cm. From 77.1 by white or yellow face sides, short eye-stripe and different range. ❋ Dense forest and scrub. ♫ Short, very high, sharp series like *Wutuwutwrwrwrwr*, last part a sharp rattle. ☉ NZ 4. E.NZ

77.3 YELLOWHEAD *Mohoua ochrocephala* [Mohoua à tête jaune] L 15cm. Unmistakable. Note thickset jizz and strong, black legs and feet. Noisy groups in canopy. ❋ Prefers tall beech forest. ♫ Very high, rapid *tsitsitsi - -* (up to 6 times). ☉ NZ 2. R E.NZ

77.4 WHITEHEAD *Mohoua albicilla* [Mohoua à tête blanche] L 15cm. From 77.3 by white head and different range. Forages at all forest levels. ❋ Tall, open, native forest. Up to 1,300m. ♫ Varied e.g. short series, starting like Chaffinch, then partly like Brown Creeper, ending like *reetseesee*. ☉ NZ 1. R E.NZ

77.5 PIPIPI (or [NZ]Brown Creeper) *Mohoua novaeseelandiae* [Mohoua pipipi] L 13cm. Note grey face sides and short, white stripe behind eye. ❋ Forest, second growth, pine plantations. ♫ Short, descending series of fluted notes, part of which inhaled *Tuut-tuweet-tuweet-Tooh*. ☉ NZ 2,3. E.NZ

77.6 GREY-STREAKED FLYCATCHER *Muscicapa griseisticta* [Gobemouche à taches grises] L 13cm. Note upright stance, thin eyering, streaked breast and narrow white margins and tips to flight feahers. ❋ Forest edge, open woodland, wooded grassland. ☉ Pa. R

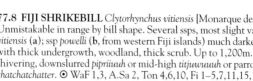

77.7 NARCISSUS FLYCATCHER *Ficedula narcissina* [Gobemouche narcisse] L 13cm. ♂♂ are unmistakable. ♀♀ and Imms are big-eyed with narrow greenish eyerings; ♀♀ have rufous tinge to uppertail coverts and tail; Imms show small but distinctive pale tips to wing coverts. ❋ Open woodland, coastal scrub, mangrove, gardens. ☉ Pa. R

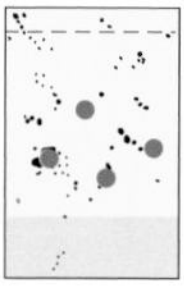

77.8 FIJI SHRIKEBILL *Clytorhynchus vitiensis* [Monarque des Fidji] L 19cm. Unmistakable in range by bill shape. Several ssps, most slight variations of ssp *vitiensis* (**a**); ssp *powelli* (**b**, from western Fiji islands) much darker above. ❋ Forest with thick undergrowth, woodland, thick scrub. Up to 1,200m. ♫ High, slightly shivering, downslurred *pipriiuuh* or mid-high *titjuwuuuh* or parrot-like, twittering *chatchatchatter*. ☉ WaF 1,3, A.Sa 2, Ton 4,6,10, Fi 1–5,7,11,15,16,18,21,22.

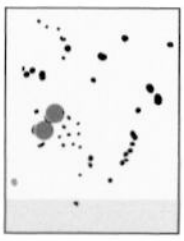

77.9 BLACK-THROATED SHRIKEBILL *Clytorhynchus nigrogularis* [Monarque à gorge noire] L 21cm. ♂ unmistakable, ♀ from 77.8 by heavier bill. ❋ Dense forest up to 1,200m. ♫ Like 77.8. ☉ Fi 1–4,11. E.Fi

78.1 RAROTONGA MONARCH *Pomarea dimidiata* [Monarque de Rarotonga] L 15cm. Unmistakable in very restricted range. Imms orange, including tail, ♂ ♂ becoming progressively grey and white in 4 years. ❋ Undergrowth of upland native forest. ♫ Rapid *tjuhwéetwéet* or *rapraprap*. ☉ Co 1,3. R E.Co

78.2 TAHITI MONARCH *Pomarea nigra* [Monarque de Tahiti] L 15cm. ♂ and ♀ similar, Imms orange. From dark morph of 76.1 by shorter grey-blue bill. ❋ Dense native forest between 80 and 400m. ♫ Powerful, toneless *kratchkratchkratch* followed by e.g. melodious *feeohweeh*. ☉ FrPo 4. R E.FrPo

78.3 MARQUESAS MONARCH *Pomarea mendozae* [Monarque des Marquises] L 17cm. Shown is sole surviving ssp *motatensis*. ♀ ♀ unmistakable. Occurs on Mohotani/Motane only. ❋ Forest and degraded forest at all elevations. ♫ Varied e.g. 2-noted *Tóh-tjep* and other, partly inhaled phrases. ☉ FrPo 28. R E.FrPo

78.4 IPHIS MONARCH *Pomarea iphis* [Monarque iphis] L 17cm. ♂ partly white below, ♀ with distinctive eyering. Occurs on Uahuka only. ❋ Forest up to 840m. Imms in dry shrub. ♫ Very high, rapid, sharp *sweesweeswee*. ☉ FrPo 13. R E.FrPo

78.5 FATUHIVA MONARCH *Pomarea whitneyi* [Monarque de Fatuhiva] L 19cm. Sexes similar. From other Marquesan monarchs by stiff, very black frontal feathers (see 78.3) or black underparts (see 78.4); only occurring on Fatuhiva, so does not share island ranges of the other species. ❋ Dense native forest and wooded thickets, 50–700m. ☉ FrPo 26. R E.FrPo

78.6 OGEA MONARCH *Mayrornis versicolor* [Monarque versicolore] L 12cm. Distinctively colour-patterned. ❋ Forest interior and edge. ☉ Fi 23,24. E.Fi

78.7 SLATY MONARCH *Mayrornis lessoni* [Monarque de Lesson] L 13cm. Note whitish lores and eyering. ❋ Dense forest, tall trees in parks and gardens. ☉ Fi thr. E.Fi

78.8 ELEPAIO *Chasiempis sandwichensis* [Monarque élépaïo] L 14cm. The five existing ssps shown: *bryani* (**a**, Ha 1), *ridgwayi* (**b**, Ha 1), Nom. (**c**, Ha 1), *ibidis* (**d**, Ha 6) and *sclateri* (**e**, Ha 7). Despite differences in plumage, especially of head, unmistakable by its cocked tail, white wing bands and white rump. ❋ Prefers closed-canopy forest above 1,100m, but can also be seen in woodland and savanna. Ssp *bryani* (most common ssp) in dry woodland at 1,800–3,100m. ♫ Twittering, rapid *witwitwit* or *wit-wit-wit* or *weetjur weetjur* or *weetjur wír*. ☉ Ha 1,6,7. E.Ha

79.1 POHNPEI FLYCATCHER *Myiagra pluto* [Monarque de Ponapé] L 15cm. No similar small, all-black flycatcher in range. Note sepia-brown breast of ♀. ✳ Forest, forest edge, open areas with scattered trees. ⊙ Mi 2. E.Mi

79.2 OCEANIC FLYCATCHER *Myiagra oceanica* [Monarque océanite] L 15cm. No similar bird in the small forest remains on Chuuk Is. ✳ Forest and forest edge. ⊙ Mi 3. E.Mi

79.3 PALAU FLYCATCHER *Myiagra erythrops* [Monarque des Palau] L 15cm. No similar small, crested bird on Palau islands. ✳ Like 79.1. ⊙ Pa 1–5. E.Pa

79.4 SATIN FLYCATCHER *Myiagra cyanoleuca* [Monarque satiné] L 17.5cm. Distinctively patterned. Note slightly peaked head and plain wings (superficially similar 85.1 has white wingbars). ✳ Migrants in open country, woodland, parks, gardens, forest clearings. ♪ Call: raspy *shrueeeh*. Song: *tchoowéeh tchoowéeh*. ⊙ NZ 1,2. V

79.5 VANIKORO FLYCATCHER *Myiagra vanikorensis* [Monarque de Vanikoro] L 13cm. From 79.7 by black bill and pale orange underparts. ✳ Forest, second growth, wooded farmland, gardens. Up to 1,100m. ♪ Hoarse, almost toneless squeaks *twittweettweet* - in rapid series of three. ⊙ Fi most islands. R E.Fi

79.6 SAMOAN FLYCATCHER *Myiagra albiventris* [Monarque des Samoa] L 14cm. No similar small bird in restricted range. ✳ Prefers multistructured forest, but also in wooded farmland and mangrove. ♪ High, tit-like *titjitjitjjitji*. ⊙ Sa. E.Sa

79.7 BLUE-CRESTED FLYCATCHER *Myiagra azureocapilla* [Monarque à crête bleue] L 16cm. Shown are Nom. (**a**, ♂ with all-black underside of tail; ♀ with white distal 2/3 of underside of tail) and ssp *castaneigularis* (**b**, ♂ and ♀ with underside of tail like ♀ **a**). Ssp *whitneyi* (not shown) like **b**, but underside of tail all-black or partly white. Red bill diagnostic. ✳ Lower levels of dense forest. ⊙ Fi 1,2,4. R E.Fi

80.1 PALAU FANTAIL *Rhipidura lepida* [Rhipidure des Palau] L 18cm. No similar bird in its range. ✳ Forest, forest remains, occasionally in mangrove. ☉ Pa 1–5. E.Pa

80.2 POHNPEI FANTAIL *Rhipidura kubaryi* [Rhipidure roux] L 16.5cm. Like 80.3 but with no or restricted rufous in plumage. No other fantail in range. ✳ Forest undergrowth. ☉ Mi 2 E.Mi

80.3 RUFOUS FANTAIL *Rhipidura rufifrons* [Rhipidure roux] L 16.5cm. Ssp *saiponensis* shown; other similar ssps found on other NMa islands and Yap. No other fantail in range. ✳ Forest undergrowth. ☉ NMa 1–4, Mi 2,4.

80.4 STREAKED FANTAIL *Rhipidura spilodera* [Rhipidure tacheté] L 17cm. Restricted to Fiji Is; shown are ssp *layardi* (**a**, Viti Levu, Ovalau) and ssp *erythronota* (**b**, N Fiji islands); not shown ssp *rufilateralis* (Taveuni, with belly deeper buff-rufous than **b**). ✳ Prefers mountain forest, but also seen in second growth and gardens. ☉ Fi 1,2,4,11,18,25.

80.5 KADAVU FANTAIL *Rhipidura personata* [Rhipidure de Kandavu] L 15cm. Only fantail on Kadavu and Ono. ✳ Forest undergrowth. ☉ Fi 3,15. E.Fi

80.6 SAMOAN FANTAIL *Rhipidura nebulosa* [Rhipidure des Samoa] L 14.5cm. Ssp *nebulosa* (**a**) restricted to Upolu and ssp *altera* (**b**) to Savaii. ✳ Dense undergrowth of forest, hedgerows, gardens. ♫ Short, very high, thin, twittered strophes *tjee-tjee-tjuh-rrrruh* or *pipirueh*. ☉ Sa. E.Sa

80.7 NEW ZEALAND FANTAIL[81] *Rhipidura fuliginosa* [Rhipidure de Nouvelle-Zélande] L 16cm. No other (arboreal) fantail in NZ. Dark morph (**a**) rather common on SI, but uncommon on NI. Amount of white in tail of pied morph (**b**) variable. ✳ All habitats with trees or scrub from forest and plantations to gardens and wooded farmland. ♫ High, unstructured series, partly clear warbling, partly nasal twittering, ending in some *tfeet* notes. ☉ NZ 1–5. E.NZ

80.8 WILLIE-WAGTAIL *Rhipidura leucophrys* [Rhipidure hochequeue] L 20cm. Forages on or near ground in open habitats. Note horizontal posture, often wagging or fanning tail. ✳ Open woodland, gardens, riverine belts, wooded grassland. ☉ NZ 4. V

80.9 SILKTAIL *Lamprolia victoriae* [Monarque queue-de-soie] L 12cm. Unmistakable by striking white rump, short tail and sparkling black plumage. Forages on the forest floor. ✳ Mature rainforest, second growth. ☉ Fi 2,4. E.Fi

81 MONARCHS & WOODSWALLOWS

81.1 TRUK MONARCH *Metabolus rugensis* [Monarque de Truk] L 20cm. ♂ unmistakable. Most ♀ ♀ retain some Imm. feathers at random through plumage. Imms bright orange with indistinct eyebrow. ✽ Undisturbed forest; also in mangroves. ☉ Mi 3. E.Mi

81.2 BLACK-FACED MONARCH *Monarcha melanopsis* [Monarque à face noire] L 18cm. Unmistakable by colour pattern. ✽ Rainforest. ☉ NZ 1. V

81.3 YAP MONARCH *Monarcha godeffroyi* [Monarque de Yap] L 15cm. Unmistakable in restricted range. ✽ Forest, second growth, scrub, mangrove. ☉Mi 4. R E.Mi

81.4 TINIAN MONARCH *Monarcha takatsukasae* [Monarque de Tinian] L 15cm. Unmistakable by contrasting white plumage parts; restricted range. ✽ Forested and wooded habitats. ☉ NMa 2. R E.NMa

81.5 WHITE-BREASTED WOODSWALLOW *Artamus leucorynchus* [Langrayen à ventre blanc] L 18cm. Note striking white rump. Sole woodswallow in range. ✽ Open areas such as savanna and pastures. ☉ Pa.

81.6 FIJI WOODSWALLOW *Artamus mentalis* [Langrayen des Fidji] L 17cm. No similar bird in range. Like all woodswallows perches openly; normally in groups. ✽ Open habitats, but occasionally seen hawking over forest. ☉ Fi 1,2,4,6,11,16. E.Fi

81.7 MASKED WOODSWALLOW *Artamus personatus* [Langrayen masqué] L 19cm. Note white-bordered face mask. ✽ Dry open woodland; occasionally in low shrubland. ♫ Series of high *tjew* notes, short chirps and rattles. ☉ NZ 2. R

81.8 WHITE-BROWED WOODSWALLOW *Artamus superciliosus* [Langrayen bridé] L 15cm. Unmistakable by colour pattern and white eyebrow. Also yellow-eyed morph shown. ✽ Open woodland, grassland with sparse shrub and trees, orchards. ♫ Series of high *tjew* notes, incorporating soft, nasal twitters. ☉ NZ 1,2. R

82 SWALLOWS, MARTINS & WHITE-EYES

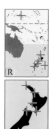

82.1 TREE MARTIN *Petrochelidon nigricans* [Hirondelle ariel] L 13cm. Note pale rump and dark brown cheeks. ✳ Occurs in a wide variety of habitats, from grassland, pastures and low shrub to woodland, forest, built-up areas. Prefers areas near wetlands. ♪ Call *sweet*. Song: twittering chatter. ☉ Mi 3, NZ 1,2,4,5. R

82.2 FAIRY MARTIN *Pterochelidon ariel* [Hirondelle ariel] L 11cm. Note striking white rump and pale chestnut head. ✳ Areas near water with sparse trees. ☉ NZ 1,2,5. R

82.3 WELCOME SWALLOW *Hirundo neoxena* [Hirondelle messagère] L 15cm. Very similar to 82.5 but not in same range. Note absence of black breast-band. ✳ Open habitats near water; common in settled areas. ♪ Song: arhythmic series of trills, nasal twitters and mewing notes. ☉ NZ 1–5,10. R

82.4 PACIFIC SWALLOW *Hirundo tahitica* [Hirondelle de Tahiti] L 13cm. The 2 ssps in the area are dark below and lack white mirrors in tail. Ssp *tahitica* (**a**, Tahiti, Moorea) darker below than ssp *subfusca* (**b**, Fiji, Tonga). ✳ Areas near rivers, waterfalls, cliffs or bridges. ♪ Very high chirps *tuweeh-tit*. ☉ Ton, FrPo 4,19, Fi.

82.5 BARN SWALLOW *Hirundo rustica* [Hirondelle rustique] L 18cm. Unmistakable by long tail streamers and white underparts. ✳ Anywhere except in forest. Often around habitation and buildings. ♪ Incessant twittering, including inhaled *sreeeh*. ☉ Ha, NMa, Pa, Mi 2–4.

82.6 ASIAN MARTIN *Delichon dasypus* [Hirondelle de Bonaparte] L 13cm. Unmistakable by neat, white-black pattern. Note white legs. ✳ Hilly country and seasides with steep slopes. ☉ Pa.

82.7 YAP WHITE-EYE *Rukia oleagineus* [Zostérops de Yap] L 13cm. Dark olive with pale yellow legs. ✳ Forest edge, woodland incl. mangrove. ☉ Mi 4. R E.Mi

82.8 TRUK WHITE-EYE *Rukia ruki* [Zostérops de Truk] L 14cm. Note orange legs. Eyering reduced to white segment under eyes. ✳ Canopy at edges of dense forest. ☉ Mi 3. E.Mi

82.9 LONG-BILLED WHITE-EYE *Rukia longirostra* [Zostérops de Ponapé] L 15cm. Note long, decurved bill and narrow eyering. ✳ Undergrowth of palm and broadleaf forest. ☉ Mi 2. R E.Mi

83.1 SAMOAN WHITE-EYE *Zosterops samoensis* [Zostérops des Samoa] L 10cm. Only white-eye on Savali. Note pale legs and eyes. ❋ Canopy of native forest and open scrub. Above 780m. ♫ Very high, soft *suhsuh* or *wuut-tjeetjee*. ☉ Sa 1. R E.Sa

83.2 LAYARD'S WHITE-EYE *Zosterops explorator* [Zostérops des Fidji] L 10cm. Restricted to Fiji. From 83.4b by green, not grey mantle. ❋ Forest, woodland, gallery forest, wooded farmland. ☉ Fi 1–4,11. R E.Fi

83.3 JAPANESE WHITE-EYE *Zosterops japonicus* [Zostérops du Japon] L 10cm. No other white-eye on Hawaiian islands. ❋ Wide variety of forested and wooded habitats. ♫ Contact call: high, thin *eee* or twittering *trutrutertrut*. Song: very high twittered phrases of any length. ☉ Ha. I

83.4 SILVER-EYE *Zosterops lateralis* [Zostérops à dos gris] L 11cm. Grey back diagnostic. No other white-eye in NZ (**a**, ssp *lateralis*); this ssp also introduced on many FrPo islands. Ssp *flaviceps* (**b**, fom Fiji) rather similar to 83.2 but with grey back. ❋ Shrub, heathland, woodland, forest, mangrove, orchards. In NZ up to 1,850m. ♫ Call: tinkling *sree*. Song: high to very high twitters such as *feefeefeetjuweeh*. ☉ FrPo 4,11,14,16,19,29, Fi, NZ.

83.5 BRIDLED WHITE-EYE *Zosterops conspicillatus* [Zostérops bridé] L 10cm. Shown are ssp *rotensis* (**a**, restricted to Rota, with yellow lores and underparts), ssp *saypani* (**b**, NMa, with white supralore) and Nom. (**c**, Guam, more yellow below than **b**, recently extirpated). ❋ Forest, thickets. ☉ NMa 1–4.

83.6 PLAIN WHITE-EYE *Zosterops hypolais* [Zostérops hypolaïs] L 10cm. Rather plain with seemingly large head. ❋ From forest canopy to grassy fields. ☉ Mi 4. E.Mi

83.7 CAROLINA ISLANDS WHITE-EYE *Zosterops semperi* [Zostérops de Semper] L 10cm. Note white supralore and all-yellow underparts. Eyes may be light chestnut. ❋ Forest, scrub. ☉ Pa 1–4, Mi 2,3.

83.8 GREY WHITE-EYE *Zosterops cinereus* [Zostérops cendré] L 10cm. Shown are Nom. (**a**, Kosrae, rather dark grey overall and often without eyering) and ssp *ponapensis* (**b**, Pohnpei, with browner upperparts than **a**). ❋ Most habitats with trees and/or shrubs. ☉ Mi 1,2. E.Mi

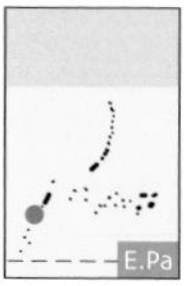

84.1 GIANT WHITE-EYE *Megazosterops palauensis* [Zostérops des Palau] L 14cm. Note yellow bill (with dusky culmen) and greyish, finely white-streaked cheeks. Cf. 75.1 with blackish eye stripe, thin bill, slimmer build and secretive habits. ❋ Forest, thickets. ⊙ Pa 3,5. R E.Pa

84.2 GOLDEN WHITE-EYE *Cleptornis marchei* [Zostérops doré] L 14cm. Unmistakable in range. ❋ Any wooded habitat. ⊙ NMa 1,3. R E.NMa

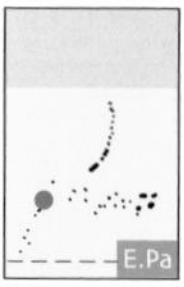

84.3 DUSKY WHITE-EYE *Zosterops finschii* [Zostérops de Finsch] L 10cm. Lacks white eyering and is darker than any other small bird on Palau Is. ❋ Any wooded habitat. ⊙ Pa 1–5. E.Pa

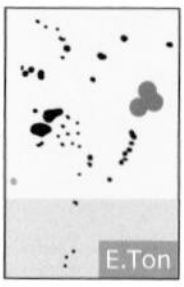

84.4 TONGAN WHISTLER *Pachycephala jacquinoti* [Siffleur des Tonga] L 18cm. ♂ unmistakable by all-black hood; ♀ like ♂, but with much paler head, darker grey on crown, pale grey on chin and throat. ❋ Undergrowth of forest, second growth. ♫ Stressed *Tjow*. ⊙ Ton. R E.Ton

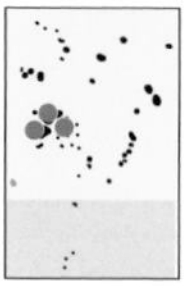

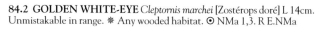

84.5 GOLDEN WHISTLER *Pachycephala pectoralis* [Siffleur doré] L 18cm. About 10 ssps in the area, ♂♂ differing mainly in colour pattern of head, ♀♀ in colouring of underparts. 2 basic ♂ types shown as examples: ssp *kandavensis* (**a**, Fi 3,9,15, with white throat; similar are *lauana*, Fi 23,26 and *vitiensis*, Fi 10) and ssp *graeffi* (**b**, Fi 1,14, with yellow throat and lacking black breast band; similar is *aurantiiventris*, Fi 2,25, with almost orange throat). Other ssps are intermediate between **a** and **b** like *optata* (**c**, Fi 1,11) and *koroana* (**d**, Fi 17). Not shown are the intermediates *ambigua* (Fi 2,18,22), *torquata* (Fi 4) and *bella* (Fi 27). ♀♀ vary between ssp **a** (♀**a**, with cinnamon underparts) and ssp **b** (♀**b**, with barred and streaked underparts). ❋ Mature forest, second growth. ⊙ Fi. R

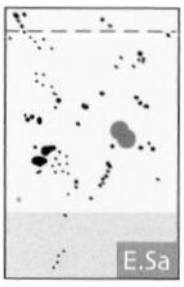

84.6 SAMOAN WHISTLER *Pachycephala flavifrons* [Siffleur des Samoa] L 17cm. Distinctive by dark grey colouring above with mottled throat. Underlying colour of ♂ throat might be pale yellow or white, mottling varies also between dense and sparse. ❋ Mature forest, second growth, gardens. ♫ Short, thrush-like strophes. ⊙ Sa. E.Sa

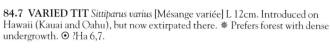

84.7 VARIED TIT *Sittiparus varius* [Mésange variée] L 12cm. Introduced on Hawaii (Kauai and Oahu), but now extirpated there. ❋ Prefers forest with dense undergrowth. ⊙ ?Ha 6,7.

85 ROBINS, MORNINGBIRD, STITCHBIRD, BELLBIRD & WATTLEBIRD

85.1 TOMTIT[82] *Petroica macrocephala* [Miro mésange] L 13cm. From larger 85.4 by white wingbar and all-white (**a**, ssp *toitoi*, NI) or yellowish (**b**, Nom, SI and Stewart I.) belly. All-black ssp *dannefaerdi* (**c**) restricted to Snares Is. Not shown ssps *chathamensis* (Chatham I.) and *marrineri* (Auckland I.), both similar to **b**. ❀ Forest, second growth, subalpine scrub, tussock grassland, pine plantations. ♫ High-pitched, simple song, slightly descending *tufec-tufec-tufec* or *teerohteerohweeh*. ☉ NZ 1–6. E.NZ

85.2 SCARLET (or [NZ]Pacific) **ROBIN**[83] *Petroica multicolor* [Miro écarlate] L 10cm. No similar small bird in range. ❀ Forest clearings and edges. Also in wooded areas and shrub. ♫ Modest *wit wit - -*. ☉ Sa, Fi 1–4.

85.3 CHATHAM (or [NZ]Black) **ROBIN** *Petroica traversi* [Miro des Chatham] L 15cm. No similar bird on Chatham I. ❀ Forest, scrub. ♫ Short, high, warbling *tutuwituwiweet-* (*wi* lower-pitched) or meandering-down *fut twéetfeet futd'r'r'r*. ☉ NZ 4. R E.NZ

85.4 NEW ZEALAND ROBIN[84] *Petroica australis* [Miro rubisole] L 18cm. Note all-blackish wings, upright stance, long legs. Shown is Nom. (ssp *rakiura*, similar to Nom., Stewart I., not shown). ❀ Forest, tall scrub, pine plantations; up to treeline. ♫ Melodious series of high to very high trills and rattles at varying speed. ☉ NZ 2–3. E.NZ

85.5 NORTH ISLAND ROBIN *Petroica longipes* [Miro de Garnot] L 18cm. Neater black than 85.4. ♀ like ♀ 85.4. ❀ Natural forest. ☉ NZ 1. E.NZ

85.6 MORNINGBIRD *Colluricincla tenebrosa* [Pitohui des Palau] L 19cm. Dull-coloured with rather large head and dark eyes. Skulker in undergrowth. ❀ Forest undergrowth. ☉ Pa 1–5. E.Pa

85.7 STITCHBIRD *Notiomystis cincta* [Méliphage hihi] L 18cm. Distinctively patterned and coloured. Note white in wings. ❀ Prefers dense native forest. ♫ Call: rapid *tsee-tih*. ☉ NZ 1,11,12 and other small islands. R E.NZ

85.8 NEW ZEALAND BELLBIRD[85] *Anthornis melanura* [Méliphage carillonneur] L 19cm. Coloured dark olive-green. Note white thin streak from bill of ♀ and Imm. down cheek. ❀ Dense forest, second growth, subalpine scrub, gardens, parks. ♫ Call: scolding *wec* or strong *tjuwtjuw*. Song: clear, rapid, liquid, ringing *pupuwéeh*. ☉ NZ 1–3,6,7. E.NZ

85.9 RED WATTLEBIRD *Anthochaera carunculata* [Méliphage barbe-rouge] L 35cm. Unmistakable if seen well. No similar bird in NZ. Note that size of wattle may vary (**a** and **b**). ❀ Prefers eucalyptus woodland, but occurs in other types of woodland and forest. Up to 1,900m. ♫ Unstructured series of short, high, parrot-like shrieks and short, toneless rattles. ☉ NZ 1. V

86.1 TUI[86] *Prosthemadera novaeseelandiae* [Tui cravate-frisée] L 30cm. Unmistakable by reflections and details in plumage. ✳ Forest, forest remains, suburbs. At lower altitudes, occasionally up to 1,500m. ♫ Very varied. Call: high, rising *huéeh*. Song: rapid, beautiful series of descending, pure, gong-like notes. ☉ NZ 1–6,10. E.NZ

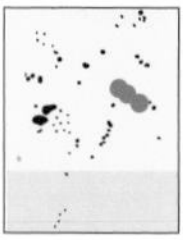

86.2 CARDINAL MYZOMELA *Myzomela cardinalis* [Myzomèle cardinal] L 12cm. Unmistakable, no similar bird in range (Samoa). ✳ Wide range of habitats incl. gardens. ♫ Very high, thin *sreeeh* and other very high, piercing shrieks and twitters. ☉ Sa, A.Sa 3.

86.3 ROTUMA MYZOMELA *Myzomela chermesina* [Myzomèle de Rotuma] L 12cm. No similar bird in range (Rotuma and nearby islets). Note black face sides of ♂♂. ✳ Wide range of habitats. ☉ Fi 21. E.Fi

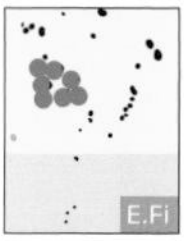

86.4 ORANGE-BREASTED MYZOMELA *Myzomela jugularis* [Myzomèle des Fidji] L 12cm. Very distinctively colour-patterned. ✳ Wide range of habitats. ☉ Fi thr. excl. 21. E.Fi

86.5 MICRONESIAN MYZOMELA *Myzomela rubratra* [Myzomèle de Micronésie] L 13cm. Note all-red head and flanks of ♂. ♀♀ differ per island: head of ssp *rubrata* shown (**a**, from Kosrae, very similar to ♂) and ssp *dichromata* (**b**, Pohnpei and nearby Ngatik). ♀♀ of ssps *saffordi* (NMa) and *major* (Chuuk Is) differ from **a** in extent of brown on mantle and intensity of red colouring; ♀♀ of ssp *kobayashii* (Palau) similar to **b**, but head more extensively suffused by red. ✳ Forest, gardens, parks. ☉ Mi 1–4, Pa, NMa.

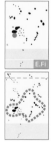

86.6 KANDAVU HONEYEATER *Xanthotis provocator* [Méliphage de Kandavu] L 18cm. Unmistakable by pattern of face sides in tiny range (Kandavu). ✳ From coastal scrub to montane forest. ☉ Fi 3. E.Fi

86.7 WATTLED HONEYEATER *Foulehaio carunculatus* [Méliphage foulehaio] L 20cm. 3 ssps, differing in pattern of face sides and dusky scalloping on underparts. Shown are Nom. (**a**, E Fi, WaF, Sa and Ton) with extensive yellow wattle and ssp *procerior* (**b**, W & C Fi), which is much darker with wattle mainly feathered black. Ssp *tavunensis* (Vanua levu, Taveuni and nearby islands) is intermediate. ✳ Coastal scrub, mangroves, montane forest. ♫ High, chattering *tjatjatja—*. ☉ WaF, A.Sa, Ton, Sa, Fi.

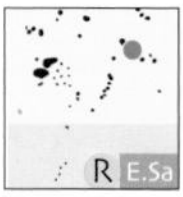

86.8 MAO *Gymnomyza samoensis* [Méliphage mao] L 28cm. Unmistakable in restricted range by large size and dark plumage. Heard more than seen. ✳ Forest, coconut plantations. ♫ Duet: *mew wjuwjutwjut*, answered by *wéetwoh* (*mew* nasal and slurred down). ☉ Sa. R E.Sa.

86.9 GIANT HONEYEATER *Gymnomyza viridis* [Méliphage vert] L 28cm. Shy, sociable. Ssp *brunneirostris* (**b**, Viti Levu) differs from Nom. (**a**, Vanua Levu and Taveuni) by dusky bill, though bill of Imm. is yellowish with brownish tip. ✳ Prefers canopy of native forest. ☉ Fi 1,2,4. R E.Fi

87 HAWAIIAN HONEYCREEPERS

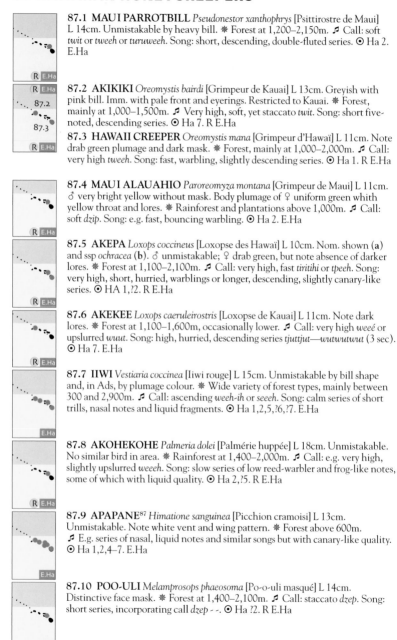

87.1 MAUI PARROTBILL *Pseudonestor xanthophrys* [Psittirostre de Maui] L 14cm. Unmistakable by heavy bill. ✳ Forest at 1,200–2,150m. ♫ Call: soft *twit* or *tweeh* or *turuweeh*. Song: short, descending, double-fluted series. ☉ Ha 2. E.Ha

87.2 AKIKIKI *Oreomystis bairdi* [Grimpeur de Kauai] L 13cm. Greyish with pink bill. Imm. with pale front and eyerings. Restricted to Kauai. ✳ Forest, mainly at 1,000–1,500m. ♫ Very high, soft, yet staccato *twit*. Song: short five-noted, descending series. ☉ Ha 7. R E.Ha

87.3 HAWAII CREEPER *Oreomystis mana* [Grimpeur d'Hawaï] L 11cm. Note drab green plumage and dark mask. ✳ Forest, mainly at 1,000–2,000m. ♫ Call: very high *tweeh*. Song: fast, warbling, slightly descending series. ☉ Ha 1. R E.Ha

87.4 MAUI ALAUAHIO *Paroreomyza montana* [Grimpeur de Maui] L 11cm. ♂ very bright yellow without mask. Body plumage of ♀ uniform green whith yellow throat and lores. ✳ Rainforest and plantations above 1,000m. ♫ Call: soft *dzip*. Song: e.g. fast, bouncing warbling. ☉ Ha 2. E.Ha

87.5 AKEPA *Loxops coccineus* [Loxopse des Hawaï] L 10cm. Nom. shown (**a**) and ssp *ochracea* (**b**). ♂ unmistakable; ♀ drab green, but note absence of darker lores. ✳ Forest at 1,100–2,100m. ♫ Call: very high, fast *tiritihi* or *tpeeh*. Song: very high, short, hurried, warblings or longer, descending, slightly canary-like series. ☉ HA 1,?2. R E.Ha

87.6 AKEKEE *Loxops caeruleirostris* [Loxopse de Kauai] L 11cm. Note dark lores. ✳ Forest at 1,100–1,600m, occasionally lower. ♫ Call: very high *weeé* or upslurred *wuut*. Song: high, hurried, descending series *tjuttjut—wutwutwut* (3 sec). ☉ Ha 7. E.Ha

87.7 IIWI *Vestiaria coccinea* [Iiwi rouge] L 15cm. Unmistakable by bill shape and, in Ads, by plumage colour. ✳ Wide variety of forest types, mainly between 300 and 2,900m. ♫ Call: ascending *weeh-ih* or *seeeh*. Song: calm series of short trills, nasal notes and liquid fragments. ☉ Ha 1,2,5,?6,?7. E.Ha

87.8 AKOHEKOHE *Palmeria dolei* [Palmérie huppée] L 18cm. Unmistakable. No similar bird in area. ✳ Rainforest at 1,400–2,000m. ♫ Call: e.g. very high, slightly upslurred *weeeh*. Song: slow series of low reed-warbler and frog-like notes, some of which with liquid quality. ☉ Ha 2,?5. R E.Ha

87.9 APAPANE[87] *Himatione sanguinea* [Picchion cramoisi] L 13cm. Unmistakable. Note white vent and wing pattern. ✳ Forest above 600m. ♫ E.g. series of nasal, liquid notes and similar songs but with canary-like quality. ☉ Ha 1,2,4–7. E.Ha

87.10 POO-ULI *Melamprosops phaeosoma* [Po-o-uli masqué] L 14cm. Distinctive face mask. ✳ Forest at 1,400–2,100m. ♫ Call: staccato *dzep*. Song: short series, incorporating call *dzep - -*. ☉ Ha ?2. R E.Ha

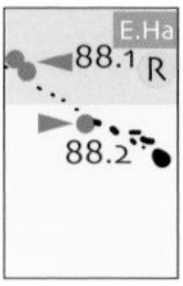

88 HAWAIIAN HONEYCREEPERS

88.1 LAYSAN FINCH *Telespiza cantans* [Psittirostre de Laysan] L 19cm. No similar bird in its tiny range. Note very heavy bill. ✳ Grass- and shrubland, the main vegetation of Laysan. ♪ High *chihchiteereh* and other carefree chirping series. ☉ Ha 13,16. R E.Ha

88.2 NIHOA FINCH *Telespiza ultima* [Psittirostre de Nihoa] L 17cm. No similar bird in its tiny range. Note heavy bill and grey plumage tones. ✳ Grass- and shrubland. ♪ Short strophes e.g. *chipchip-chéeroh* or *tuweréeh* or chirping series. ☉ Ha 8. R E.Ha

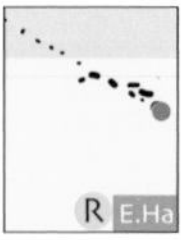

88.3 OU *Psittirostra psittacea* [Psittirostre psittacin] L 17cm. Note pink, heavy, hooked bill. Yellow head of ♂ diagnostic. ✳ Wet forest at 800–1,900m. ♪ Call: rising *uiih* or descending *piuuh*. Song: rich flow of high, melodious notes, most of which repeated 3–4 times. ☉ Ha 1. R (probably extinct) E.Ha

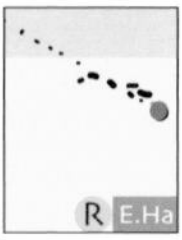

88.4 PALILA *Loxioides bailleui* [Psittirostre palila] L 19cm. Plumage pattern of yellow and grey tones diagnostic. ✳ Dry forest at 2,000–3,000m. ♪ Call: *ptweet, ohweet* and variations thereof. Song: series of modest, high *teetjuw* and other notes with many short repetitions. ☉ Ha 1. R E.Ha

88.5 NUKUPUU *Hemignathus lucidus* [Hémignathe nukupuu] L 14cm. Unmistakable in range. Note curved lower mandible. ✳ Dense forest at 1,450–2,000m (Maui) or 1,000–1,300m (Kauai). ☉ ?Ha (probably extinct). R E.Ha

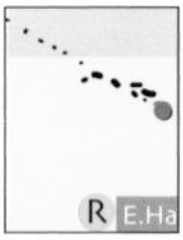

88.6 AKIAPOLAAU *Hemignathus munroi* [Hémignathe akiapolaau] L 14cm. From 88.5 by range and straight lower mandible. ✳ Forest and parklands. ♪ Call: *tuwjur* and *tuweéh*. Song: loud *wutwutwididjur*, last part louder and higher in pitch. ☉ Ha 1. R E.Ha

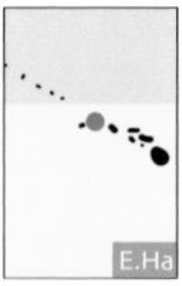

88.7 ANIANIAU *Hemignathus parvus* [Petit Amakihi] L 10cm. 88.7–10 very similar, differing mainly in shape and colour of bill. Range diagnostic, except for 88.7 and 88.10. This species restricted to Kauai. Pink legs and bill diagnostic. ✳ Native forest, mainly above 600m. ♪ Call: upslurred *tweet* or very high, short *pfit*. Song: varied e.g. short, high, almost level warblings with tit-like variations. ☉ Ha 7. E.Ha

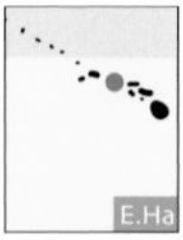

88.8 OAHU AMAKIHI *Hemignathus flavus* [Amakihi d'Ohau] L 11cm. This species restricted to Oahu. Short, almost straight bicoloured bill. Note wingbars of ♀. ✳ Native forest; also in plantations and trees in urban areas. ☉ Ha 6. E.Ha

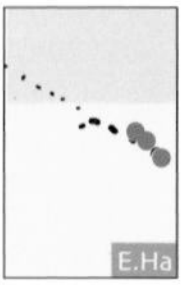

88.9 HAWAII AMAKIHI *Hemignathus virens* [Amakihi d'Hawaï] L 11cm. Note curved, dusky bill. ♀ ♀ variable between and within ssps in plumage colour (some much greener, others much greyer). Nom. (**a**) restricted to Hawaii, ssp *wilsoni* (**b**) to Mauai and Molokai. Note wingbars of ♀ **b**. ✳ Rather dry forest. ♪ Song: high, rhythmic, descending *tweet tweet - - tweet* and other songs with similar structure and repetition of the same note or phrase. ☉ Ha 1,2,5. E.Ha

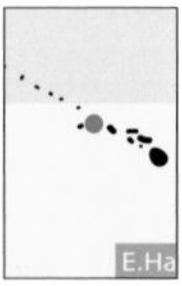

88.10 KAUAI AMAKIHI *Hemignathus kauaiensis* [Amakihi de Kauai] L 11cm. Note rather heavy bill (not pink as in 88.7). Found on Kauai. ✳ Forest above 600m. ♪ Call: *weeéh* or nasal *wew*. High, 3-noted, slightly descending *tjeutjewtjew* or reed-warbler-like series. ☉ Ha 7. E.Ha

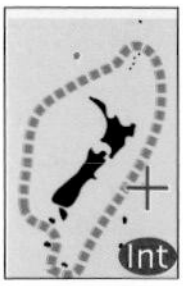

89.1 YELLOWHAMMER *Emberiza citrinella* [Bruant jaune] L 16cm. From 89.2 by paler central crown patch, greener mantle and buffier rump. ✳ Farmland with hedgerows, open pastures with some trees and thickets, roadsides, parks and gardens. ♫ Short series, 1st part very high, sizzling, almost rattling *sisisisisi*, 2nd part lower-pitched *feeeh*. ☉ NZ 1–5,7,10. I

89.2 CIRL BUNTING *Emberiza cirlus* [Bruant zizi] L 16cm. See 89.1. ✳ Farmland with some trees and hedgerows, willow-fringed wetlands, occasionally in parks and gardens. ♫ Like 89.1, but lacking 2nd part. ☉ NZ 1,2. I

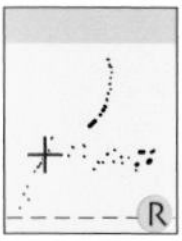

89.3 BLACK-HEADED BUNTING *Emberiza melanocephala* [Bruant mélanocéphale] L 16.5cm. ♂ is unmistakable. ♀ is grey with pale yellow vent. ✳ Prefers farmland and grassland. ♫ Soft *tsutsu*, immediately followed by loud, descending *prupuperwéer*. ☉ Pa. V

89.4 SNOW BUNTING *Plectrophenax nivalis* [Bruant des neiges] L 17cm. In all plumages with complete or partly white secondaries. In flight with broad white outer tail edges and white rump, forming an inverted V that contrasts with black central tail feathers. ✳ Prefers seashore and coastal grassland. ☉ Ha 11,17,18. V

89.5 SAVANNAH SPARROW *Passerculus sandwichensis* [Bruant des prés] L 14cm. Note yellowish eyebrows, large, pink feet and streaked plumage. ✳ Normally found in grassland and marsh. ♫ Extreme high, thin *tsip*. ☉ Ha 11,18. V

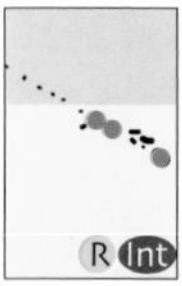

89.6 SAFFRON FINCH *Sicalis flaveola* [Sicale bouton-d'or] L 14cm. Normally seen in short grass. Shown are ♂ (**a**) and 1st W ♂ (**b**) *brasiliensis* and ♀ *pelzelni* (**c**). ✳ Semi-open areas with scattered bush and trees and short grass, such as lawns and golf courses. ♫ Very/ultra high, unstructured, warbling *tit-weertit- -tjitweet* (2 sec). ☉ Ha 1,?2,6,7. I

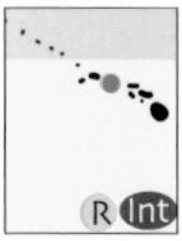

89.7 YELLOW-FACED GRASSQUIT *Tiaris olivaceus* [Sporophile grand-chanteur] L 11cm. Unmistakable by small yellow bib and eyebrow. ✳ Weedy and grassy areas, thickets, roadsides. ♫ Soft, rather toneless rattle. ☉ Ha 6. I

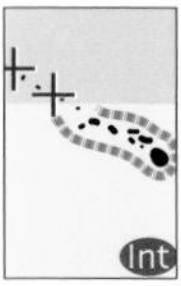

89.8 HOUSE FINCH *Carpodacus mexicanus* [Roselin familier] L 15cm. ♀♀ are rather nondescript and streaky; ♂♂ occur with much red (**a**), orange or yellow (**b**) in head, throat and chest. ✳ Wide variety of habitats from towns to forest and open woodland. ♫ *wurk*; also e.g. descending *witwitwitwittefjeetohfjeet*. ☉ Ha 1–8,11,17. I

89.9 OLIVE-BACKED ORIOLE *Oriolus sagittatus* [Loriot sagittal] L 28cm. Vagrant from Australia to SI. Note size, slender build and red eye. ✳ Prefers edges of clearings in eucalyptus forest and woodland, but also found in mangroves and riverine belts. ☉ NZ 2. V

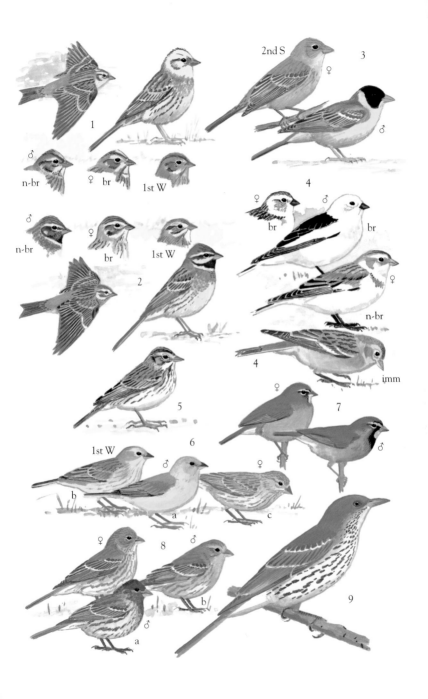

1

♂ n-br   ♀ br   1st W

2nd S   ♀   3

♂   4

♂ n-br   ♀ br   1st W

2

♀ br   ♂ br

♀ n-br

4   imm

5

♀   7

♂

1st W   6

b   ♂   ♀

a   c

♀   8   ♂

b

a   9

90 CHAFFINCH, GREENFINCH, GOLDFINCH, REDPOLL, CANARIES & SPARROWS

90.1 CHAFFINCH *Fringilla coelebs* [Pinson des arbres] L 14.5cm. Striking white wingbar and shoulder diagnostic. Note green rump. ✳ Farmland, forest, parks, gardens, tussock grassland, weedy areas, occasionally on coastal mudflats and beaches. ♫ Call: high, upslurred *wéeh*. Song: energetic, lowered *tsjiptsjiptsjip-biskweeh*. ⊙ NZ. I

90.2 EUROPEAN GREENFINCH *Carduelis chloris* [Verdier d'Europe] L 16cm. Note compact size and heavy bill. Shows striking yellow wingbar and tail sides, especially in flight. ✳ Habitats with exotic plants, such as parks, gardens, farmland, plantations, dunes, shrubland. ♫ Call: nasal, drawn-out *sreeeh* or short twitters. ⊙ NZ 1–5,7,10. I

90.3 EUROPEAN GOLDFINCH *Carduelis carduelis* [Chardonneret élégant] L 12cm. Unmistakable by colour pattern. Imm. shows pale grey head, but also the yellow wingbar as in Ad. ✳ Habitats with trees and hedgerows esp. those with thistles, like farmland, roadsides, forest clearings, wasteland. ♫ Song: fast, liquid, mellow warbling. ⊙ NZ 1–8,10. I

90.4 COMMON REDPOLL *Carduelis flammea* [Sizerin flammé] L 12cm. Red front and black bib diagnostic. ✳ Open areas, farmland, parks, dense shrubland, forest clearings. ♫ Call: questioning, drawn-up *pueeét*. ⊙ NZ, Ha 17,18. I V

90.5 ISLAND (or ^{AOU}Common) **CANARY** *Serinus canaria* [Serin des Canaries] L 15cm. Most common seen is the all-yellow form with whitish wings (see thumbnail); the ancestor of the domestic canary, as shown, does not occur in the area. ✳ Tree stands, open woodland, gardens, cultivation. ⊙ Ha 17. I

90.6 YELLOW-FRONTED CANARY *Serinus mozambicus* [Serin du Mozambique] L 11cm. Note the distinctive facial pattern with diagnostic black malar. ✳ Open (occasionally dense) forest, parks. ♫ Very high *tu-wéet* or thin, rather sharp, sustained warbling. ⊙ Ha 1,6. I

90.7 HOUSE SPARROW *Passer domesticus* [Moineau domestique] L 15cm. Well known. ♂♂ have black bill in Br plumage. ✳ Mainly near human habitation. ♫ Well-known chirps. ⊙ Ma 1, NZ 1–8,10, Ha. I

90.8 EURASIAN TREE SPARROW *Passer montanus* [Moineau friquet] L 14cm. From larger 90.7 by red-brown crown, white cheeks and black ear patch. ✳ Mainly near human habitation. ♫ Like 90.7 but slightly higher pitched. ⊙ Ma 1, Gu, Mi 4, Pa 2,5,11, NMa 1,2,4. I

91.1 LAVENDER WAXBILL *Estrilda caerulescens* [Astrild queue-de-vinaigre] L 11cm. From 91.2 by red tail and black, white-spotted lower flanks. ✴ Agricultural fields, parks, roadsides and other weedy, grassy areas. ♫ Very high *twit* and *twitwutwit*. ☉ Ha 1,6. I

91.2 BLACK-TAILED WAXBILL *Estrilda perreini* [Astrild à queue noire] L 11cm. See 91.1. ✴ Like 91.1. ☉ ?Ha 6. I

91.3 ORANGE-CHEEKED WAXBILL *Estrilda melpoda* [Astrild à joues orange] L 11cm. Unmistakable by orange-yellow ear coverts. Note yellow tinge to belly. ✴ Like 91.1. ♫ Mid-high, nasal twittering. ☉ NMa 1, Ha 2,6. I

91.4 BLACK-RUMPED WAXBILL *Estrilda troglodytes* [Astrild cendré] L 10cm. Note neat, black, white-rimmed tail. ✴ Like 91.1. ♫ Very high *féeeh*. ☉ Ha 1,?6. I

91.5 COMMON WAXBILL *Estrilda astrild* [Astrild ondulé] L 10cm. Body plumage overall delicately barred. Undertail coverts black, those of Imm. not so. ✴ Like 91.1. ♫ Call: high, short chirps or *snee*. Song: short, inhaled phrase *tuhfeetutuh*. ☉ FrPo 4,19, Ha ?6. I

91.6 BLUE-FACED PARROTFINCH *Erythrura trichroa* [Diamant de Kittlitz] L 11.5cm. Unmistakable by blue face sides. ✴ Rainforest, mangrove, scrub, grassland, open pastures. ☉ Pa 1–5, Mi 1–3.

91.7 PINK-BILLED PARROTFINCH *Erythrura kleinschmidti* [Diamant à bec rose] L 11cm. Unmistakable by large, strong bill and short tail. ✴ Mainly in mature, wet forest up to 1,000m. ☉ Fi 1. E.Fi

91.8 FIJI PARROTFINCH[88] *Erythrura pealii* [Diamant de Peale] L 10cm. ♂ and ♀ similar (mask of ♀ slightly more orange). Note black chin and black line bordering lower red cheek. ✴ Grassland, rice fields, parks, gardens. ☉ Fi 1–4,6,28. E.Fi

91.9 RED-HEADED PARROTFINCH *Erythrura cyaneovirens* [Diamant vert-bleu] L 10cm. Colours of underparts less saturated than 91.8 and black on chin and around face mask missing. ✴ Rainforest, second growth, plantations. ☉ Sa.

92 CORDONBLEU, AVADAVAT, FIRETAIL, SILVERBILL & MUNIAS

92.1 RED-CHEEKED CORDONBLEU *Uraeginthus bengalus* [Cordonbleu à joues rouges] L 13cm. Unmistakable. No similar bird in the area. ✳ Weedy areas such as road sides, neglected culture, over-grazed fields. ♫ Call: very high *swee*. ☉ Ha 1. I

92.2 RED AVADAVAT *Amandava amandava* [Bengali rouge] L 11cm. Unmistakable by white dots on red and brown plumage. ✳ Like 92.1. ♫ Long series of high *weet* notes, interspersed with small inhalations, e.g. *feet*. ☉ Fi 1–5, Ha 2,6. I

92.3 RED-BROWED FIRETAIL *Aegintha temporalis* [Diamant à cinq couleurs] L 11cm. No bird with similar eyebrow and dark grey plumage in the area. ✳ Like 92.1. ☉ FrPo 4,12,13,19,20,23. I

92.4 AFRICAN SILVERBILL *Euodice cantans* [Capucin bec-d'argent] L 11cm. From Imms 92.5–9 by black tail, very pale plumage and faint barring of wing coverts and back. ✳ Like 92.1, but mainly in drier areas. ♫ Very high, thin *weetohweet*. ☉ Ha 1–7. I

92.5 NUTMEG MANNIKIN *Lonchura punctulata* [Capucin damier] L 11cm. Note scalloping to underparts of Ad. Imms with black bill and mantle concolorous with rest of upperparts. ✳ Like 92.1. ♫ Mid-high to high *feet*. ☉ Mi 4, Ha 1–7. I

92.6 MOTTLED MUNIA *Lonchura hunsteini* [Capucin de Hunstein] L 11cm. Ads unmistakable. Imm. is rather dark tawny. ✳ Like 92.1. ♫ ☉ Mi 2,?3. I

92.7 BLACK-HEADED (or ^AOU^Tricolored) **MUNIA** *Lonchura malacca* [Capucin à dos marron] L 11cm. Ads unmistakable. Imms rather uniform brownish with pale grey bill. ✳ Like 92.1. ♫ High, dry *wutwut wut*. ☉ Ha 6,7. I

92.8 CHESTNUT-BREASTED MUNIA *Lonchura castaneothorax* [Capucin donacole] L 11cm. From other munias by range. Ads unmistakable. ✳ Like 92.1. ☉ FrPo 4,14,16,19,20,25,28. I

92.9 CHESTNUT MUNIA *Lonchura atricapilla* [Capucin à tête noire] L 11cm. Like 92.7 but underparts concolorous with upperparts. ✳ Like 92.1. ☉ Ha 6,7, Gu, Pa 1,2,5. I

93 CARDINALS, TANAGER, MEADOWLARK, GRACKLE & DRONGO

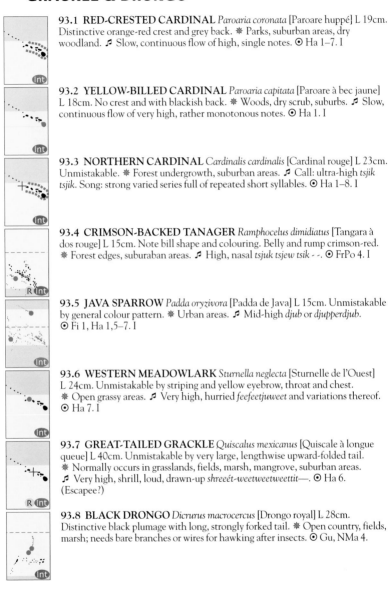

93.1 RED-CRESTED CARDINAL *Paroaria coronata* [Paroare huppé] L 19cm. Distinctive orange-red crest and grey back. ✳ Parks, suburban areas, dry woodland. ♫ Slow, continuous flow of high, single notes. ☉ Ha 1–7. I

93.2 YELLOW-BILLED CARDINAL *Paroaria capitata* [Paroare à bec jaune] L 18cm. No crest and with blackish back. ✳ Woods, dry scrub, suburbs. ♫ Slow, continuous flow of very high, rather monotonous notes. ☉ Ha 1. I

93.3 NORTHERN CARDINAL *Cardinalis cardinalis* [Cardinal rouge] L 23cm. Unmistakable. ✳ Forest undergrowth, suburban areas. ♫ Call: ultra-high *tsjik tsjik*. Song: strong varied series full of repeated short syllables. ☉ Ha 1–8. I

93.4 CRIMSON-BACKED TANAGER *Ramphocelus dimidiatus* [Tangara à dos rouge] L 15cm. Note bill shape and colouring. Belly and rump crimson-red. ✳ Forest edges, suburaban areas. ♫ High, nasal *tsjuk tsjew tsik - -*. ☉ FrPo 4. I

93.5 JAVA SPARROW *Padda oryzivora* [Padda de Java] L 15cm. Unmistakable by general colour pattern. ✳ Urban areas. ♫ Mid-high *djub* or *djupperdjub*. ☉ Fi 1, Ha 1,5–7. I

93.6 WESTERN MEADOWLARK *Sturnella neglecta* [Sturnelle de l'Ouest] L 24cm. Unmistakable by striping and yellow eyebrow, throat and chest. ✳ Open grassy areas. ♫ Very high, hurried *feefeetjuweet* and variations thereof. ☉ Ha 7. I

93.7 GREAT-TAILED GRACKLE *Quiscalus mexicanus* [Quiscale à longue queue] L 40cm. Unmistakable by very large, lengthwise upward-folded tail. ✳ Normally occurs in grasslands, fields, marsh, mangrove, suburban areas. ♫ Very high, shrill, loud, drawn-up *shreeét-weetweetweettit—*. ☉ Ha 6. (Escapee?)

93.8 BLACK DRONGO *Dicrurus macrocercus* [Drongo royal] L 28cm. Distinctive black plumage with long, strongly forked tail. ✳ Open country, fields, marsh; needs bare branches or wires for hawking after insects. ☉ Gu, NMa 4.

94 STARLINGS & MYNAS

94.1 SAMOAN STARLING *Aplonis atrifusca* [Stourne de Samoa] L 30cm. Very dark overall, almost black below. ❋ Forest clearings, plantations, coastal fringe, gardens. ♫ High *tju-wrrreh* (last part slightly rising). ☉ Sa, A.Sa.

94.2 RAROTONGA STARLING *Aplonis cinerascens* [Stourne de Rarotonga] L 21cm. Mouse-grey, mottled paler. ❋ Native forest at 150–600m. ♫ Varied: e.g. melodious, well-spaced notes such as *prrruh*, *wrihwrih* and *rrreeet*. ☉ R Co 1. E.Co

94.3 POLYNESIAN STARLING *Aplonis tabuensis* [Stourne de Polynésie] L 19cm. 10 ssps, differing in darkness of plumage, striping below and whiteness of edges to secondaries. Shown: *vitiensis* (**a**, Fi), Nom. (**b**, SE Fi and Ton) and *tutuilae* (**c**, Tutuila). Eyes normally yellow on E Fi islands, but brown (**d**) on W Fi Is. ❋ Mainly at edge and near clearings of forest, on smaller islands in any wooded habitat. ♫ Rather toneless *tjuh-wrueeeh* and loud, sharp *tiew-tiew-tiew* or *weet-weet-weet*. ☉ WaF, Ni, Ton 6,7,10, A.Sa, Sa, Fi.

94.4 MICRONESIAN STARLING *Aplonis opaca* [Stourne de Micronésie] L 24cm. Black plumage and yellow eyes diagnostic. ❋ Varied habitats at all elevations. ☉ NMa, Pa, Mi.

94.5 POHNPEI STARLING *Aplonis pelzelni* [Stourne de Pohnpei] 16cm. Very small; plumage is sooty black without reflections. Note dark eyes. ❋ Montane forest. ☉ Mi 2. E.Mi

94.6 WHITE-CHEEKED STARLING *Sturnus cineraceus* [Étourneau gris] L 22cm. Distinctive white cheeks (irregularly striped dark brown), rump and edges of secondaries. ❋ Farmland, open woodland, parks, towns. ☉ NMa 1. R

94.7 JUNGLE MYNA *Acridotheres fuscus* [Martin forestier] L 23cm. Crest and lack of yellow face skin diagnostic. In flight like 94.8. ❋ Wooded farmland, parks, gardens. ♫ Short, high, loud, slightly thrush-like phrases. ☉ Ton 7, A.Sa 3, Fi 1,2,4,8,11,29,30, Sa. I

94.8 COMMON MYNA *Acridotheres tristis* [Martin triste] L 25cm. Distinctive facial pattern with pale blue eyes. From 94.7 by bare, yellow skin to face sides. ❋ In and near areas of human habitation. ♫ Incessant stream of short strophes, composed by clear, mewing, harsh, gurgling notes, each repeated 2–4 times. ☉ NZ 1, Ma, FrPo 20, Sa, A.Sa 3, Co, Fi, Ha. I

94.9 HILL MYNA *Gracula religiosa* [Mainate religieux] L 29cm. Unmistakable. Note heavy legs. ❋ Could occur in parks (last seen in Lyon Arboretum). ☉ ?Ha 6. I

95 STARLINGS, MAGPIE, KOKAKO, SADDLEBACK & CROWS

95.1 EUROPEAN STARLING *Sturnus vulgaris* [Étourneau sansonnet] L 21cm. Spotted appearance distinctive. Does not hop, but walks on lawns and such. Imm. brown, attaining Ad. plumage gradually, the head staying pale brown longest. ❋ Wide variety of habitats, esp. human habitation and farmland, less in woodland, rarely in forest. ♪ Incessant stream of varied, clear, hoarse, rattling, croaking, chirping and dry-rattled notes, accompanied by wing-clapping. ☉ Ton 2,11,12, ?Fi, Ha 1,6, NZ. I

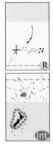

95.2 CHESTNUT-CHEEKED STARLING *Sturnia philippensis* [Étourneau à joues marron] L 17cm. Unmistakable by colour pattern, ♀ ♀ showing a bleached-out version. ❋ Normally winters in open country and towns. ☉ Pa. R

95.3 AUSTRALASIAN (or ᴺᶻAustralian) **MAGPIE** *Gymnorhina tibicen* [Cassican flûteur] L 41cm. Unmistakable by size, colour pattern, bill shape and noisy behaviour; **a** (white-backed) and **b** (black-backed) are colour forms. ❋ Open, grassy or partly bare habitats with some scattered trees and hedgerows. Also in bordering forest or woodland. ♪ Short series of melodious, gurgling, fluted and guttural notes. ☉ NZ, Fi 4,8,29,31. I

95.4 KOKAKO[89] *Callaeas cinereus* [Glaucope cendré] L 38cm. No similar bird in NZ. Shown are Nom. (**a**, SI, with yellow-orange wattle) and ssp *wilsoni* (**b**, with blue wattle). ❋ Mainly in forests up to 900m. ♪ Beautiful, calm stream of well-separated, loud, fluted notes of varying pitch. ☉ NZ 1,?2,?3. R E.NZ

95.5 SADDLEBACK[90] *Philesturnus carunculatus* [Créadion rounoir] L 25cm. No similar bird in its range. ❋ Forest, coastal and montane shrubland. ♪ E.g. nasal, shivering and chattering series or high-pitched, sneezing *tseeh* notes. ☉ NZ, restricted to small islands off mainland. R E.NZ

95.6 ROOK *Corvus frugilegus* [Corbeau freux] L 45cm. Only crow in NZ. Note white base of Ad. bill, which in Imm. is feathered and with nasal bristle. ❋ Farmland with stands of high trees. ♪ Low, raspy *furaah*. ☉ NZ 1,2,4. I

95.7 HAWAIIAN CROW *Corvus hawaiiensis* [Corneille d'Hawaï] L 49cm. No other crow in Ha, but probably not surviving in the wild. ❋ Forest, woodland, grassy areas with scattered trees. ♪ Very varied, e.g. high, raucous, slightly upslurred *Aoow* or falsetto, trumpet-like *Rrruah*. Generally higher pitched than other crow species. ☉ Ha 1. R E.Ha

95.8 MARIANA CROW *Corvus kubaryi* [Corneille de Guam] L 40cm. Only crow in Guam and Rota. Note slender bill. ❋ Forest. ☉ Gu, NMa 4. R

ENDNOTES

THE FOLLOWING NOTES SERVE TWO PURPOSES:

1. To refer to the systematics and nomenclature agreed on by the Ornithological Society of New Zealand (OSNZ) for New Zealand and by the American Ornithological Union (AOU) for Hawaii. In general it concerns names of regional species and subspecies and splits in species and subspecies not found in James F. Clements *The Clements Checklist of the Birds of the World*, 6th Edition (Helm, 2007). The differences between species and subspecies split by the OSNZ often seem to be determined by range and not by features that are distinctively visible in the field; therefore, the concerned species and subspecies are not illustrated on the plates or mentioned in the captions and their names are only given in the following notes. Likewise subspecies that are mentioned by Clements but not by the OSNZ (because they occur elsewhere, outside New Zealand) are not illustrated or mentioned in the captions or notes, unless they are identifiable in the field. The regional English names of species and subspecies on the lists produced by the OSNZ and by the AOU are added to the index of this book.

2. To provide some extra information (e.g. on identification) that, because of available space, could not be inserted in the accounts opposite the plates.

[1] **1.7 AUSTRALASIAN** or [NZ]Eastern Little **GREBE** *Tachybaptus novaehollandiae*
Nom. [NZ]Australasian Little Grebe

[2] **1.8 GREAT CRESTED GREBE** *Podiceps cristatus*
ssp *australis* [NZ]Australasian Crested Grebe

[3] **PLATE 2** The colour of the kiwi plumages as shown might differ from the colour as seen in the field; not only is there some variation among individuals, but also the colour might appear different when seen at night in torch light, when most kiwis would normally be seen. The brown kiwis (2.1–3) are easily distinguished from the spotted kiwis (2.4–5) by their striped or barred plumage. The best way to identify species is by range.

[4] **2.3 SOUTHERN BROWN KIWI** *Apteryx australis*
Nom. [NZ]South Island Brown Kiwi
ssp *lawryi* [NZ]Stewart Island Brown Kiwi

[5] **2.8 GENTOO PENGUIN** *Pygoscelis papua*
Nom. [NZ]Northern Gentoo Penguin

[6] **3.4 ROCKHOPPER PENGUIN** *Eudyptes chrysocome*
The OSNZ splits this species into 3 species:
[NZ]EASTERN ROCKHOPPER PENGUIN *Eudyptes filholi*
[NZ]MOSELEY'S ROCKHOPPER PENGUIN *Eudyptes moseleyi*
[NZ]WESTERN ROCKHOPPER PENGUIN *Eudyptes chrysocome*

[7] **4.1 WANDERING ALBATROSS** *Diomedea exulans*
Clements recognises 5 ssps of which 3 can be seen in the area (sometimes at the same moment); might be identified safely only when seen at their nest on breeding grounds: Nom. (**a**, tends to become the whitest, especially ♂♂, breeds Inaccessible and Gough Is.), *gibsoni* (**b**, intermediate, retaining Imm. stages 3–5, breeds Antipodes and Campbell Is.) and

antipodensis (**c**, tends to stay the darkest, especially ♀♀, which can retain Juv. and early Imm. stages 5–7 during life; breeds Auckland I.). [NZ] ssp *chionoptera* [NZ]Snowy Albatross (not mentioned by Clements), ressembling Nom. is also no longer recognised by the OSNZ as a valid ssp and merged into Nom. The OSNZ treats *gibsoni* and *antipodensis* as ssps of an independent species ANTIPODEAN ALBATROSS *Diomedea antipodensis*.

[8] **4.2 ROYAL ALBATROSS** *Diomedea epomophora*
The OSNZ treats the 2 ssps as independent species:
[NZ]NORTHERN ROYAL ALBATROSS *Diomedea sanfordi*
[NZ]SOUTHERN ROYAL ALBATROSS *Diomedea epomophora*

[9] **6.2 YELLOW-NOSED ALBATROSS**
The OSNZ treats the 2 subspecies as independent species:
[NZ]ATLANTIC YELLOW-NOSED ALBATROSS *Thalassarche chlororhynchos*
[NZ]INDIAN OCEAN YELLOW-NOSED ALBATROSS *Thalassarche carteri/bassi*

[10] **6.3 BLACK-BROWED ALBATROSS**
The OSNZ treats the two ssps as independent species:
BLACK-BROWED ALBATROSS *Thalassarche melanophris*
[NZ]CAMPBELL BLACK-BROWED ALBATROSS *Thalassarche impavida*

[11] **6.4 SHY** or [NZ]White-capped **ALBATROSS** *Thalassarche cauta*
The OSNZ splits this species into 3 species:
[NZ]WHITE-CAPPED ALBATROSS *Thalassarche cauta*
Nom. [NZ]Tasmanian Albatross
ssp *steadi* [NZ]New Zealand White-capped Albatross
[NZ]CHATHAM ISLAND ALBATROSS

Thalassarche eremita
[NZ]SALVIN'S ALBATROSS *Thalassarche salvini*

[12] **6.5 BULLER'S ALBATROSS** *Thalassarche bulleri*
Nom. [NZ]Southern Buller's Albatross
ssp *platei* [NZ]Northern Buller's Albatross

[13] **8.6 FULMAR PRION** *Pachyptila crassirostris*
Nom. Fulmar Prion
ssp *flemingi* [NZ]Lesser Fulmar Prion
ssp *pyramidalis* [NZ]Chatham Fulmar Prion

[14] **9.3 GOULD'S PETREL** *Pterodroma leucoptera*
In NZ: ssp *caledonica* [NZ]New Caledonian Petrel

[15] **9.6 GREAT-WINGED PETREL** *Pterodroma macroptera*
Nom. Great-winged Petrel
ssp *gouldi* [NZ]Grey-faced Petrel

[16] **12.1 CAPE PETREL** *Daption capense*
Nom. Cape Petrel
ssp *australe* [NZ]Snares Cape Petrel

[17] **12.2 SNOW PETREL** *Pagodroma nivea*
Nom. [NZ]Lesser Snow Petrel
ssp *major* [NZ]Greater Snow Petrel

[18] **14.7 MANX SHEARWATER** *Puffinus puffinus*
The OSNZ splits this species in two species:
MANX SHEARWATER *Puffinus puffinus*
[NZ]NEWELL'S SHEARWATER *Puffinus newelli*

[19] **14.8 LITTLE SHEARWATER** *Puffinus assimilis*
Nom. [NZ]Norfolk Island Little Shearwater
ssp *haurakensis* [NZ]North Island Little Shearwater
ssp *kermadecensis* [NZ]Kermadec Little Shearwater
ssp *elegans*: The OSNZ rises this ssp to independent species:
[NZ]SUBANTARCTIC LITTLE SHEARWATER *Puffinus elegans*

[20] **15.5 WHITE-FACED STORM-PETREL** *Pelagodroma marina*
ssp *dulciae* [NZ]Australian White-faced Storm-Petrel
ssp *maoriana* [NZ]New Zealand White-faced Storm-Petrel
The OSNZ rises *albiclunis* to independent species:
[NZ]KERMADEC STORM-PETREL *Pelagodroma albiclunis*

[21] **16.8 COMMON DIVING-PETREL** *Pelecanoides urinatrix*
Nom. [NZ]Northern Diving-Petrel
ssp *chathamensis* [NZ]Southern Diving-Petrel
ssp *exsul* [NZ]Subantarctic Diving-Petrel

[22] **18.4 DARTER** or [AOU]Anhinga *Anhinga melanogaster*
ssp *novaehollandiae* [NZ]Australian Darter

[23] **18.5 GREAT CORMORANT** *Phalacrocorax carbo*
ssp *novaehollandiae* [NZ]Black Shag

[24] **18.7 LITTLE PIED CORMORANT** *Microcarbo melanoleucos*
Nom. Little Pied Cormorant
ssp *brevirostris* [NZ]Little Shag

[25] **PLATE 19** Many shags show a complicated sequence of plumages; from N-br plumage they moult into a prebreeding plumage, which changes during incubating into Br plumage. A good example is **19.3 ROUGH-FACED SHAG**: the prebreeding plumage shows an orange swelling at the bill base, which shrinks during incubating without the bird losing its crest; the crest is lost after breeding.

[26] **19.1 SPOTTED SHAG** *Phalacrocorax punctatus*
Nom. Spotted Shag
ssp *oliveri* [NZ]Blue Shag

[27] **19.2–19.8** is a group of shags ressembling each other closely; they are best identified on basis of range. Note also the differences in facial pattern, especially of bare parts. 19.2, 19.4 and 19.7 have a white dorsal patch. Shown are full body of Br birds and heads of N-br and Imm. birds. All rest and nest on rocky shores or on ledges, ridges and tops of steep cliffs, from where they forage nearby or far out at sea.

[28] **20.3 EASTERN GREAT EGRET** *Ardea modesta*
Species level follows Kushlan and Hancock (2005).

[29] **20.10 CATTLE EGRET** *Bubulcus ibis*
ssp *coromanda* [NZ]Eastern Cattle Egret

[30] **21.2 RUFOUS NIGHT-HERON** *Nycticorax caledonicus*
ssp *australasiae* [NZ]Nankeen Night Heron

[31] **21.7 LITTLE BITTERN** *Ixobrychus minutus*
ssp *dubius* [NZ]Australian Little Bittern

[32] **22.1 GREY HERON** *Ardea cinerea*
ssp *jouyi* [NZ]Oriental Grey Heron

[33] **22.7 AUSTRALIAN IBIS** *Threskiornis molucca*
ssp *strictipennis* [NZ]Australian White Ibis

[34] **FULVOUS WHISTLING-DUCK** *Dendrocygna bicolor* was possibly introduced on Oahu and Kauai, but is now considered extirpated. Resembles **23.3 PLUMED WHISTLING-DUCK** but lacks flank-barring and shows smaller flank plumes.

[35] **25.3 AUSTRALIAN** or [NZ]Australasian **SHOVELER** *Anas rhynchotis*
ssp *variegata* [NZ]New Zealand Shoveler

[36] **26.5 GREEN-WINGED TEAL** *Anas carolinensis*. By AOU treated as ssp of **26.4 EURASIAN TEAL**

[37] The OSNZ has added **WHITE-BELLIED SEA EAGLE** *Haliaeetus leucogaster* to the NZ checklist on basis of a single specimen in the Museum of New Zealand Te Papa Tongarewa. Not depicted, no map provided. Slightly smaller than Steller's and White-tailed Sea Eagle; underparts, head and most of tail white; wings and back grey.

[38] The OSNZ has added **CORNCRAKE** *Crex crex* to the NZ checklist on basis of a single specimen in the Museum of New Zealand Te Papa Tongarewa. Not depicted, no map provided. Size about that of Plain Bush-Hen. Greyish brown with dark spots above, lengthwise arranged in rows, grey throat and upper breast and disinctive reddish bars at flanks.

[39] **36.7 BUFF-BANDED RAIL** *Gallirallus philippensis*
ssp *assimilis* NZBanded Rail
ssp *mellori* NZBuff-banded Rail

[40] **36.8 WEKA** *Gallirallus australis*
Nom. NZWestern Weka, found mainly in the N and W regions of SI. Distinguished by dark red-brown and black streaking on the breast. The Western Weka has 3 distinct colour phases, that of the southernmost range showing more black;
ssp *greyi* NZNorth Island Weka, found in Northland and Poverty Bay and released elsewhere. This ssp differs in its greyer underparts, and brown rather than reddish-coloured legs;
ssp *hectori* NZBuff Weka, confined to Chatham I. and Pitt I. Has a lighter overall colouring than the other ssps:
ssp *scotti* NZStewart Island Weka is smaller than the other ssps and has 2 colour phases; a chestnut form (similar to the chestnut-phase Western Weka) and a black phase which is not as dark as the black Western Weka. The population is confined to Stewart I./Rakiura. Taxonomic status of Weka on Kapiti I. is unclear.

[41] **36.9 AUCKLAND ISLANDS RAIL** *Lewinia muelleri*. Sometimes treated as ssp of *Lewinia pectoralis* LEWIN'S RAIL from Melanesia.

[42] The OSNZ has added **AUSTRALIAN CRAKE** *Porzana fluminea* to the NZ checklist on basis of a single specimen in the Museum of New Zealand Te Papa Tongarewa. Not depicted, no map provided. Size about that of other crakes; resembles 37.1 Baillon's Crake, but grey plumage parts much darker and with red base to upper mandible.

[43] **37.1 BAILLON'S CRAKE** *Porzana pusilla*
ssp *affinis* NZMarsh Crake
ssp *palustris* NZBaillon's Crake

[44] **38.5 PURPLE SWAMPHEN** *Porphyrio porphyrio*
ssp *samoensis*
ssp *pelewensis*
ssp *melanotus*
The OSNZ treats ssp *P. p. melanotus* as NZPukeko NZPorphyria melanotus melanotus, which is then Nom. of NZSOUTH-WEST PACIFIC SWAMPHEN NZPorphyria melanotus

[45] **38.9 EURASIAN COOT** *Fulica atra*
ssp *australis* NZAustralian Coot

[46] **39.4 SOUTH ISLAND** NZPied **OYSTERCATCHER** *Haematopus finschi*
Treated by OSNZ as ssp of **39.4 SOUTH ISLAND OYSTERCATCHER**. Interbreeds in E coastal areas of SI with **39.1 VARIABLE OYSTERCATCHER**, resulting in hybrids, which resemble 39.4 but with smudgier features.

[47] **39.6 PIED STILT** *Himantopus leucocephalus*
Treated by OSNZ as ssp of **39.5 BLACK-WINGED STILT**. 39.6 PIED STILT and

39.7 BLACK STILT interbreed freely, resulting in a wide range of hybrids (some examples shown).

[48] **40.4 MASKED LAPWING** *Vanellus miles*
ssp *novaehollandiae* NZSpur-winged Plover

[49] **42.3 DOUBLE-BANDED PLOVER** *Charadrius bicinctus*
Nom. NZBanded Dotterel
ssp *exilis* NZAuckland Island Banded Dotterel

[50] **42.5 RED-BREASTED** or NZNew Zealand **DOTTEREL** *Charadrius obscurus*
Nom. NZSouthern New Zealand Dotterel
ssp *aquilonius* NZNorthern New Zealand Dotterel

[51] **43.1 BLACK-TAILED GODWIT** *Limosa limosa*
ssp *melanuroides* NZAsiatic Black-tailed Godwit

[52] **43.2 BAR-TAILED GODWIT** *Limosa lapponica*
ssp *baueri* NZEastern Bar-tailed Godwit

[53] **43.7 WHIMBREL** *Numenius phaeopus*
ssp *hudsonicus* NZAmerican Whimbrel
ssp *variegatus* NZAsiatic Whimbrel

[54] **46.2 SUBANTARCTIC SNIPE** *Coenocorypha aucklandica*
The OSNZ splits this species in 2 full species:
NZSNARES ISLAND SNIPE *Coenocorypha huegeli*; and
SUBANTARCTIC SNIPE *Coenocorypha aucklandica* with 2 ssps:
Nom. NZAuckland Island Snipe
ssp *meinertzhagenae* NZAntipodes Island Snipe

[55] **48.1 BROAD-BILLED SANDPIPER** *Limicola falcinellus*
ssp *sibirica* NZEastern Broad-billed Sandpiper

[56] **48.2 RED KNOT** *Calidris canutus*
ssp *rogersi* NZLesser Knot

[57] **50.1 BROWN** or NZSouthern **SKUA** *Stercorarius antarcticus*
ssp *lonnbergi* NZSubantarctic Skua

[58] **51.3 RED-BILLED GULL** *Larus scopulinus*
HANZAB and OSNZ merge 51.3 and 51.4 and recognise only RED-BILLED GULL (not SILVER GULL) *Larus novaehollandiae* with ssp *scopulinus* in NZ.

[59] **52.1 KELP GULL** *Larus dominicanus*
Nom. NZSouthern Black-backed Gull

[60] **52.4 AMERICAN HERRING GULL** *Larus smithsonianus*
Treated by AOU as ssp of HERRING GULL *L. argentatus*

[61] **54.1 FAIRY TERN** *Sternula nereis*
ssp *davisae* NZNew Zealand Fairy Tern

[62] **54.3 LITTLE TERN** *Sternula albifrons*
ssp *sinensis* NZEastern Little Tern

[63] **54.4 COMMON TERN** *Sterna hirundo*
ssp *longipennis* NZEastern Common Tern

[64] **54.6 ANTARCTIC TERN** *Sterna vittata*
ssp *bethunei* NZNew Zealand Antarctic Tern

[65] **55.5 GREY NODDY** *Procelsterna albivitta*
By OSNZ treated as subspecies of **55.6 BLUE-**NZgrey **NODDY** *Procelsterna cerulea*

⁶⁶ **55.6 BLUE-** ^{NZ}grey **NODDY** *Procelsterna cerulea*
ssp *albivitta* ^{NZ}Grey Noddy

⁶⁷ **58.1 NEW ZEALAND PIGEON** *Hemiphaga novaeseelandiae*
By OSNZ split in 2 species:
NEW ZEALAND PIGEON *Hemiphaga novaeseelandiae*
^{NZ}CHATHAM ISLAND PIGEON *Hemiphaga chathamensis*

⁶⁸ By OSNZ listed as BARBARY DOVE *Streptopelia risoria*

⁶⁹ **61.3 RED SHINING-PARROT** *Prosopeia tabuensis*
Systematics unclear: at least 2 ssps exist (Clements), probably more (BotW).
Birds introduced to Gau (Fi) and Eua (Ton) are now believed to be the result of interbreeding of at least 2 ssps.

⁷⁰ **63.7 NEW ZEALAND KAKA** *Nestor meridionalis*
Nom. ^{NZ}South Island Kaka
ssp *septentrionalis* ^{NZ}North Island Kaka

⁷¹ **64.3 RED-FRONTED** or ^{NZ}-crowned **PARAKEET** *Cyanoramphus novaezelandiae*
Split by OSNZ into 2 species, the first with 3 ssps:
^{NZ}RED-CROWNED PARAKEET *Cyanoramphus novaezelandiae*
Nom. ^{NZ}Red-crowned Parakeet
ssp *chathamensis* ^{NZ}Chatham Island Red-crowned Parakeet
ssp *cyanurus* ^{NZ}Kermadec Parakeet;
^{NZ}REISCHEK'S PARAKEET *Cyanoramphus hochstetteri*

⁷² **65.9 SHINING BRONZE-CUCKOO** *Chrysococcyx lucidus*
Nom. ^{NZ}Shining Cuckoo

⁷³ **67.3 SACRED KINGFISHER** *Todiramphus sanctus*
ssp *norfolkiensis* ^{NZ}Norfolk Island Kingfisher
ssp *vagans* ^{NZ}New Zealand Kingfisher

⁷⁴ **69.1 BARN OWL** *Tyto alba*
Taxonomy of Barn Owls is under discussion.
ssp *delicatula* ^{NZ}Australian Barn Owl and *furcata*, a group of New World ssps among which *pratincola* [I to Hawaii] might be independent species.

⁷⁵ **69.5 MOREPORK** *Ninox novaeseelandiae*
Nom. Morepork
ssp *undulata* ^{NZ}Norfolk Island Boobook

⁷⁶ **70.8 AUSTRALASIAN** or ^{NZ}New Zealand **PIPIT** *Anthus novaeseelandiae*
Nom. ^{NZ}New Zealand Pipit
ssp *aucklandicus* ^{NZ}Auckland Island Pipit
ssp *chathamensis* ^{NZ}Chatham Island Pipit
ssp *steindachneri* ^{NZ}Antipodes Island Pipit

⁷⁷ **71.1 RIFLEMAN** *Acanthisitta chloris*
Nom. ^{NZ}South Island Rifleman
ssp *granti* ^{NZ}North Island Rifleman

⁷⁸ **74.9 FERNBIRD** *Megalurus punctatus*
Nom. ^{NZ}South Island Fernbird
ssp *caudata* ^{NZ}Snares Island Fernbird

ssp *stewartiana* ^{NZ}Stewart Island Fernbird
ssp *vealeae* ^{NZ}North Island Fernbird
ssp *wilsoni* ^{NZ}Codfish Island Fernbird

⁷⁹ **PLATE 76** *Acrocephlus* reed-warbler species are widespread thr. the Pacific area. Many show 'leucism' (partial albinism, but no red eyes) in regular, symmetric plumage patterns and/or irregular distributed feathers or feather groups. Most tend to become whiter after each moult. 76.1 occurs also as a rare melanistic morph.

⁸⁰ **77.2 CHATHAM ISLAND GERYGONE** or ^{NZ}Warbler *Gerygone albofrontata*
By the OSNZ treated as ssp of **77.1 GREY GERYGONE**

⁸¹ **80.7 NEW ZEALAND FANTAIL** *Rhipidura fuliginosa*
ssp *placabilis* ^{NZ}North Island Fantail
Nom. ^{NZ}South Island Fantail
ssp *penita* ^{NZ}Chatham Island Fantail

⁸² **85.1 TOMTIT** *Petroica macrocephala*
ssp *toitoi* ^{NZ}North Island Tomtit
Nom. ^{NZ}South Island Tomtit
ssp *chathamensis* ^{NZ}Chatham Island Tomtit
ssp *dannefaerdi* ^{NZ}Snares Island Tomtit
ssp *marrineri* ^{NZ}Auckland Island Tomtit

⁸³ **85.2 SCARLET** or ^{NZ}Pacific **ROBIN** *Petroica multicolor*
Nom. ^{NZ}Norfolk Island Robin

⁸⁴ **85.4 NEW ZEALAND ROBIN** *Petroica australis*
The OSNZ merges **85.4 NEW ZEALAND ROBIN** *Petroica australis* and **85.5 NORTH ISLAND ROBIN** *Petroica longipes* into 1 species **NEW ZEALAND ROBIN** with 3 ssps:
Nom. ^{NZ}South Island Robin
ssp *longipes* ^{NZ}North Island Robin
ssp *rakiura* ^{NZ}Stewart Island Robin

⁸⁵ **85.8 NEW ZEALAND BELLBIRD** *Anthornis melanura*
Nom. ^{NZ}Bellbird
ssp *obscura* ^{NZ}Three Kings Bellbird
ssp *oneho* ^{NZ}Poor Knights Bellbird

⁸⁶ **86.1 TUI** *Prosthemadera novaeseelandiae*
Nom. Tui
ssp *chathamensis* ^{NZ}Chatham Island Tui

⁸⁷ **87.9 APAPANE** *Himatione sanguinea*
Note: ssp *freethi* (Laysan) extinct.

⁸⁸ **91.8 FIJI PARROTFINCH** *Erythrura pealii*
91.8 is often treated as ssp of **91.9 RED-HEADED PARROTFINCH**

⁸⁹ **95.4 KOKAKO** *Callaeas cinereus*
By OSNZ split into 2 independent species:
Callaeas cinereus ^{NZ}SOUTH ISLAND KOKAKO
Callaeas wilsoni ^{NZ}NORTH ISLAND KOKAKO

⁹⁰ **95.5 SADDLEBACK** *Philesturnus carunculatus*
By OSNZ split into 2 independent species:
Philesturnus carunculatus ^{NZ}NORTH ISLAND SADDLEBACK
Philesturnus rufusater ^{NZ}SOUTH ISLAND SADDLEBACK

NATIONAL AND INTERNATIONAL ORGANISATIONS

The Ornithological Society of New Zealand

www.osnz.org.nz/index.html

Aims and objectives:

- To encourage, organise and promote the study of birds and their habitat use, particularly within the New Zealand region.
- To foster and support the wider knowledge and enjoyment of birds generally.
- To promote the recording and wide circulation of the results of bird studies and observations.
- To produce a journal and any other publication containing matters of ornithological interest.
- To effect cooperation and exchange of information with other organisations having similar aims and objectives.
- To assist the conservation and management of birds by providing information, from which sound management decisions can be derived.
- To maintain a library of ornithological literature for the use of members and to promote a wider knowledge of birds.
- To promote the archiving of observations, studies and records of birds, particularly in the New Zealand region.
- To carry out any other activity that is capable of being conveniently carried out in connection with the above objects, or that directly or indirectly advances those objects or any of them.

La Société d'Ornithologie de Polynésie 'MANU'

www.manu.pf/F_Lasop.html

La Société d'Ornithologie de Polynésie 'MANU', association sans but lucratif, est une organisation non gouvernementale fondée en juillet 1990 par quelques amateurs passionnés par les oiseaux de Polynésie Française. 'MANU' est affiliée à 'BirdLife International', fédération mondiale d'associations œuvrant pour la conservation des oiseaux. 'MANU' est membre de la Fédération des Associations de Protection de la Nature 'Te Ora Naho'.

Les buts de 'MANU':

- La protection des oiseaux de Polynésie et de leurs habitats.
- La contribution à l'étude des oiseaux de Polynésie dans leur milieu naturel.
- La diffusion et la promotion auprès du public de toute information relative à la protection et à l'étude des oiseaux de Polynésie.

Les activités de 'MANU':
- Publication du bulletin de liaison 'Te Manu' (4 numéros par an), destinés aux membres.
- Participation à des actions éducatives et de sensibilisation du public (Journée de l'environnement, Conférences, Radio, Télévision).
- Sorties sur le terrain (observations, enregistrements sonores, photographies...) et accueil et soutien logistique de missions ornithologiques en Polynésie Française.
- Propositions de projets de programmes de protection, financement et réalisation d'actions de suivi et de protection des espèces menacées.

Government of New Zealand

www.doc.govt.nz

This Department of Conservation (DOC) site has information about the protection of New Zealand's natural and historic heritage, how and where you can enjoy public conservation places and how to get involved in conservation.

AOU The American Ornithologists' Union

www.aou.org

Founded in 1883, the American Ornithologists' Union is the oldest and largest organisation in the New World devoted to the scientific study of birds. Although the AOU primarily is a professional organisation, its membership of about 4,000 includes many amateurs dedicated to the advancement of ornithological science. The Union publishes *The Auk*, a quarterly journal of ornithology that publishes papers that deal with ornithological research.

NatureServe

www.natureserve.org/index.jsp

NatureServe is a organisation that provides on a non-profit base scientific information for conservation action in the USA, Canada, Latin America and the Caribbean.

BirdLife International

www.birdlife.org

BirdLife International is a global partnership of conservation organisations that strives to conserve birds, their habitats and global biodiversity, working with people towards sustainability in the use of natural resources. BirdLife Partners operate in over 100 countries and territories worldwide.

Fatbirder

www.fatbirder.com

Fatbirder is the premier web resource about birds, birding and birdwatching for birders – hundreds of pages and tens of thousands of links about birding everywhere in the world: a page for every country and state; every bird family; books, guides, forums, reserves, accommodation, trip reports, bird clubs, ornithology, twitching, conservation, optics, holidays. So if you want to go birding, or want to know about particular birds, then Fatbirder is for you! Each page has sections on useful reading, guides, mailing lists, reserves, places to stay, trip reports, bird clubs, etc. This website benefits hugely from the contributions of birders, photographers and ornithologists from all over the world. Most of the country and state introductions are written by, and most photographs featured taken by, Fatbirder users.

BIBLIOGRAPHY AND FURTHER REFERENCES

Field Guides

Moon, G. 2002. *Photographic Guide to Birds of New Zealand*. New Holland Publishers.

Pratt, H.D., Jr, Bruner, P.L. and Berrett, D.G. 1989. *A Field Guide to the Birds of Hawaii and the Tropical Pacific*. Princeton University Press.

Robertson, H. and Heather, B. 1999. *The Hand Guide to the Birds of New Zealand*. Oxford University Press.

Shirihai, H. 2002. *The Complete Guide to Antarctic Wildlife*. Oxford University Press.

Watling, D. 2004. *A Guide to the Birds of Fiji & Western Polynesia: Including American Samoa, Niue, Samoa, Tokelau, Tonga, Tuvalu and Wallis & Futuna*. Environmental Consultants, Fiji.

Field Guides for Adjacent and Other Regions

Coates, B.J. and Peckover, W.S. 2001. *Birds of New Guinea and the Bismarck Archipelago: A Photographic Guide*. Dove Publications.

Doughty, C., Day, N. and Plant, A.1999. *Birds of The Solomons, Vanuatu and New Caledonia*. Christopher Helm.

Flegg, J. (text). 2002. *Birds of Australia: Photographic Field Guide* [Most photographs supplied by Nature Focus at the Australian Museum]. Frenchs Forest.

Kennedy, R., Gonzales, P.C., Dickinson, E., Miranda, H. and Fisher, T. 2000 *A Guide to the Birds of the Philippines*. Oxford University Press.

Mullarney, K., Lars Svensson, L., Zetterstrom, D. and Grant, P. 1999. *Collins Bird Guide*. Collins.

Shimba, T. 2009. *A Photographic Guide to the Birds of Japan and North-East Asia*. Christopher Helm.

Sibley, D. 2000. *The North American Bird Guide*. Pica Press.

Simpson, K. and Day, N. 1996. *Field Guide to the Birds of Australia*. Viking Publishers.

Slater, P., Slater, P. and Slater, R. 1989. *The Slater Field Guide to Australian Birds*. Landsdowne Publishers.

Other Basic Publications

The Checklist Committee (Dr Brian Gill, Convener) Ornithological Society of New Zealand. 2010. *Checklist of the Birds of New Zealand and the Ross Dependency, Antarctica*. Te Papa Press in association with the Ornithological Society of New Zealand Inc., 4th edn.

Thornton, J. 2009. *The Field Guide to New Zealand Geology*. Penguin Books, NZ.

Christidis, L. and Boles, W.E. 2008. *Systematics and Taxonomy of Australian Birds*. CSIRO Publishing.

Kushlan, J.A. and Hancock, J.A. 2005 *Herons*. Oxford University Press.

Handbooks

Handbook of Australian, New Zealand and Antarctic Birds (Oxford University Press):

(Vol. 1) *Ratites to Ducks*. Marchant, S. and Higgins, P.J. (eds) 1990.

(Vol. 2) *Raptors to Lapwings*. Marchant, S. and Higgins, P.J. (eds) 1993.

(Vol. 3) *Snipe to Pigeons*. Higgins, P.J. and Davies, S.J.J.F. (eds) 1996. (Vol. 4) *Parrots to Dollarbird*. Higgins, P.J. (ed.) 1999.

(Vol. 5) *Tyrant-flycatchers to Chats*. Higgins, P.J., Peter, J.M. and Steele, W.K. (eds) 2001.

(Vol. 6) *Pardalotes to Shrike-thrushes*. Higgins, P.J. and Peter, J.M. (eds) 2002.

(Vol. 7) *Boatbill to Starlings*. Higgins, P.J., Peter, J.M. and Cowling, S.J. (eds) 2006.

J. del Hoyo, Elliott, A. and Sargatal, J. (eds). *Handbook of the Birds of the World*. Lynx Editions, Barcelona.

(Vol. 1) *Ostrich–Ducks*. 1992.

(Vol. 2) *New World Vultures–Guineafowl*. 1994.

(Vol. 3) *Hoatzin–Auks*. 1996.

(Vol. 4) *Barn-owls–Hummingbirds*. 1997.

(Vol. 5) *Sandgrouse–Cuckoos*. 1999.

(Vol. 6) *Mousebirds–Hornbills*. 2001.

(Vol. 7) *Jacamars–Woodpeckers*. 2002.

(Vol. 8) *Broadbills–Tapaculos*. 2003.

(Vol. 9) *Cotingas–Pipits and Wagtails*. 2004.

(Vol. 10) *Cuckoo-shrikes–Thrushes*. 2005.

(Vol. 11) *Old World Flycatchers–Old World Warblers*. 2006.

(Vol. 12) *Picathartes–Tits and Chickadees*. 2007.

(Vol. 13) *Penduline-Tits–Shrikes*. 2008.

(Vol. 14) *Bush-Shrikes–Old World sparrows*. 2008.

Volumes 15 and 16 yet to be published.

Books on Bird Groups

Brooke, M. and Cox, J. 2004. *Albatrosses and Petrels Across the World*. Oxford University Press.

Byers, C., Olsson, U. and Curson, J. 1995. *Buntings and Sparrows*. Pica Press.

Chantler, P. and Driessens, G. 1995. *Swifts*. Pica Press.

Cleere, N. and Nurney, D. 1998. *Nightjars*. Pica Press.

Clement, P., Harris, A. and Davis, J. 1993. *Finches and Sparrows*. Helm.

Clement, P. and Hathway, R. 2000. *Thrushes*. Helm.

Enticott, J. and Tipling, D. 1997. *Photographic Handbook of the Seabirds of the World*. New Holland Publishers Ltd.

Ferguson-Lees, J. and Christie, D.A. 2001. *Raptors of the World*. Helm.

Fry, C.H., Fry, K. and Harris, A. 1992. *Kingfishers, Bee-eaters and Rollers*. Helm.

Gibbs, D. Barnes, E. and Cox, J. 2001. *Pigeons and Doves*. Pica Press.

Hancock, J. and Kushlan, J. 1984. *The Herons Handbook*. Croom Helm.

Harrison, P. 1983. *Seabirds*. Croom Helm.

Hayman, P., Marchant, J. and Prater, T. 1986. *Shorebirds*. Helm.

Isler, M.L. and Isler, P.R. 1999. *Tanagers*. Helm.

Jaramillo, A. and Burke, P. 1999. *New*

World Blackbirds. Helm.

Jupiter, A. and Parr, M. 1998. *Parrots.* Pica Press.

König, C., Weick, F. and Becking, J.-H. 1999. *Owls.* Pica Press, Helm.

Madge, S. and Burn, H. 1988. *Wildfowl.* Helm.

Madge, S. and Burn, H. 1999. *Crows and Jays.* Helm.

Olsen, K.M. and Larsson, H. 1997. *Skuas and Jaegers.* Pica Press.

Taylor, B. and van Perlo, B. 1998. *Rails.* Pica Press.

Turner, A. and Rose, C. 1998. *Swallows and Martins.* Helm.

Bird Sounds

Cornell Laboratory of Ornithology/ Interactive Audio. *Peterson Field Guides, Western Bird Songs.*

McPherson, L. 2002. *New Zealand Birds – A Sound Guide*, Vols 1 to 7. McPherson Natural History Unit.

Vol. 1: *Kiwis to Blue Petrel.*

Vol. 2: *Broad-billed Prion to Pitt Island Shag.*

Vol. 3: *Darter to Brolga.*

Vol. 4: *Banded Rail to Turnstone.*

Vol. 5: *Chatham Island Snipe to Kea.*

Vol. 6: *Crimson Rosella to Auckland Island Tit.*

Vol. 7: *North Island Robin to Rook.*

McPherson, L. 2002. *Birds of Polynesia.* McPherson Natural History Unit.

McPherson, L. 2002. *More Birds of Polynesia.* McPherson Natural History Unit.

Pratt, H.D. 1996. *Voices of Hawaii's Birds* [Two cassette tapes with accompanying text]. Hawaii Audubon Society, Honolulu, and Cornell Laboratory of Ornithology, Ithaca, New York.

The Internet

Information and photos of almost every bird species can be found on the Internet via its English, French or scientific name.

www.rosssilcock.com (A Birder's Checklist of the Birds of the Pacific Region)

www.pwlf.org/birds.htm (The Pacific WildLife Foundation – Birds of the Pacific – Science, Behaviour, Distribution, Description, Photos and Video)

www.digimages.info/listeoiseauxmonde/ intro_cinfo.htm (la Commission Internationale des Noms Français des Oiseaux, CINFO)

www.aou.org/checklist/north/index.php (AOU – Check-list of North American Birds)

www.npwrc.usgs.gov/resource/birds/ chekbird/ r1/hawaiias.htm (Checklist Hawaii)

www.oiseaux.net/oiseaux/polynesie. francaise.famille.html (Les oiseaux de Polynésie française)

www.idahobirds.net/identification/white-cheeked/subspecies.html (Subspecies Accounts for Cackling Goose and Canada Goose)

www.osqar.se/arter/Utlandet/Nya%20 Zeeland/Galleri (Photos of birds New Zealand)

www.kapitiislandalive.co.nz/index.html (Lodging Kapiti Island)

www.albatrossencounter.co.nz/albatross/ ocean_wings (Seabird excursions)

www.arthurgrosset.com (Photos of neotropical and other birds)

www.mangoverde.com/birdsound/index. html (Photos of birds worldwide)

www.hbw.com/ibc (Video clips of birds worldwide)

APPENDIX

Species that have become extinct since the beginning of the 19th century:

Akioloa *Hemignathus obscurus* 1940

Amaui *Myadestes woahensis* <1860

Auckland Island Merganser *Mergus australis* 1902

Barred-winged Rail *Nesoclopeus poecilopterus* 1973?

Bishop's O-o *Moho bishopi* 1915

Black Mamo *Drepanis funerea* 1907

Black-fronted (or Tahiti) Parakeet *Cyanoramphus zealandicus* ca.1850

Bush Wren *Xenicus longipes* 1972

Chatham Island Rail *Cabalus* (or *Rallus*) *modestus* ca.1900

Chatham Islands Fernbird *Megalurus* (*Bowdleria*) *rufescent* <1900

Dieffenbach's Rail *Gallirallus dieffenbachii* mid-1900

Eiao Flycatcher *Pomarea fluxa* ca.1977

Greater Akialoa *Hemignathus ellisianus* ca.1940

Greater Amakihi *Hemignathus sagittirostris* 1901

Greater Koa Finch *Rhodacanthis palmeri* 1895

Guam Flycatcher *Myiagra freycineti* 1983

Hawaii Mamo *Drepanis pacifica* ca.1898

Hawaii O-o *Moho nobilis* 1934

Hawaiian Crake *Porzana sandwichensis* 1890

Hawkins's Rail *Diaphorapteryx hawkinsi* 1895?

Hiua *Heteralocha acutirostris* ca.1922

Kakawahie *Paroreomyza flammea* 1963

Kaua'i 'O'O (or O-o) *Moho braccatus* 1987

Kioea *Chaetoptila angustipluma* ca.1859

Kona Grosbeak *Chloridops kona* 1894

Kosrae Crake *Porzana monasa* ca.1875

Kosrae Starling *Aplonis corvina* <1931

Lanai Hookbill *Dysmorodrepanis munroi* 1918

Laughing Owl *Sceloglaux albifacies* 1914

Laysan Crake *Porzana palmeri* 1944

Lesser Akialoa *Hemignathus obscurus* 1940

Lesser Koa Finch *Rhodacanthis flaviceps* 1891

Lyall's (or Stephen's Island) Wren *Xenicus lyalli* 1895

Maupiti Flycatcher *Pomarea pomarea* ca.1825

Mira Flycatcher *Pomarea mira* ca.1990

Moorean Sandpiper *Prosobonia ellisi* <1900

Mysterious Starling *Aplonis mavornata* ca.1850

New Zealand Bittern *Ixobrychus novaezelandiae* 1890

New Zealand Quail *Coturnix novaezelandiae* 1875

North Island Piopio *Turnagra tanagra* ca.1885

Northern Takahe *Porphyrio mantelli* 1894

Nuku hiva Flycatcher *Pomarea nukuhivae* 1930?

Nuku-pu'u *Hemignathus lucidus* 2000?

Oahu (Alauahio or) Creeper *Paroreomyza maculata* 1990?

Oahu O-o *Moho apicalis* 1837

Pohnpei Starling *Aplonis pelzelni* Last seen in 1994

Raiatea Parakeet *Cyanoramphus ulietanus* <1800?

Red-moustached Fruit-Dove *Ptilinopus mercierii* 1980?

Robust White-eye *Zosterops strenuus* ca.1925

Samoan Moorhen *Gallinula* (or *Pareudiastes*) *pacifica* 1870?

South Island Piopio *Turnagra capensis* 1905

Tahiti Rail *Gallirallus pacificus* ca.1900

Ula-ai-hawane *Ciridops anna* 1937?

Wake Rail *Gallirallus wakensis* 1945?

White-winged (or Tahitian) Sandpiper *Prosobonia leucoptera* <1900

INDEX

The names are printed as in the species descriptions, English names in **BOLD CAPITALS**, French in regular script and scientific names in *italics*.

254